TALES FROM THE
UNDERGROUND

DAVID W. WOLFE

TALES FROM THE
UNDERGROUND

*A Natural History
of Subterranean Life*

PERSEUS PUBLISHING
Cambridge, Massachusetts

Cataloging-in-Publication Data is available from the Library of Congress
ISBN 0-7382-0128-6

Perseus Publishing is a member of the Perseus Books Group.

Find us on the World Wide Web at http://www.perseuspublishing.com

Perseus Publishing books are available at special discounts for bulk purchases in the U.S. by corporations, institutions, and other organizations. For more information, please contact the Special Markets Department at the Perseus Books Group, 11 Cambridge Center, MA 02142, or call 617-252-5298.

Text design by Jeffrey P. Williams
Set in 11-point Minion by Perseus Publishing Services

First printing, April 2001

1 2 3 4 5 6 7 8 9 10—03 02 01

Contents

1755

TO MY FAMILY

Acknowledgments

I am pleased to have an opportunity to thank publicly the many friends, family members, and colleagues who have contributed to this book. I doubt that I would have had the stamina to complete this project without the unwavering and enthusiastic support of Terry Kristensen, who also carefully read through the earliest, sometimes long and tedious, drafts of each chapter and provided many comments that greatly improved the book's tone and readability. I am very grateful to my editor, Amanda Cook, who believed early on that these "tales" were worth telling and who championed the project. In addition to providing many detailed suggestions that improved the prose, Amanda's skillful guidance regarding the structure of individual chapters and the book as a whole was invaluable. I would also like to thank the others at Perseus Books who have contributed in various ways, and Tamara Clark for her wonderful illustrations.

This book is broad in scope, encompassing soil science, microbiology, biogeochemistry, ecology, human and plant pathology, animal behavior, genetics, and evolutionary biology. To tackle all of this, I called on many of my scientist colleagues for help, particularly Dean Biggins, Tom Eisner, Bill Ghiorse, Gary Harman, Bob Howarth, George Hudler, Ken Mudge, Aly Naguib, Janice Thies, Lynne Trulio, and Carl Woese. I thank them for being so very generous with their time, openly sharing their expertise and personal experiences, and reviewing portions of the manuscript. I was not able to incorporate all of their suggestions and had to make difficult decisions regarding what to leave in and what to leave out. The responsibility for any serious omissions or errors in the final product is completely mine.

I feel fortunate to work at an institution—Cornell University—that has a long history of encouraging faculty to communicate scientific information to the general public. I benefited from access to the outstanding collection of resources maintained at Cornell's Mann Library and from the exceptional service of the dedicated staff who work there. Some of this book was written while on a sabbatical leave at the University of Nevada at Reno. I would like to thank Jeff Seemann of the Department of Biochemistry for providing me with a comfortable environment in which to work while I was there.

Special thanks to Diane Ackerman for some useful advice early on about writing and publishing. Thanks to Alan and Laura Falk for letting me use their lake house as a hideaway at a time when I most needed it. Finally, I thank all of my friends, my daughter, Alexis, and other family members for their cheerful understanding during long periods of neglect as I completed this ambitious project.

TALES FROM THE
UNDERGROUND

Introduction

*Man and man's earth are unexhausted
and undiscovered.
Wake and listen!
Verily, the earth shall yet be a source of recovery.*
—FRIEDRICH NIETZCHE (1911)
AND THUS SPAKE ZARATHUSTRA

ONE DOESN'T HAVE TO VENTURE FAR INTO THE UNDERGROUND for new discoveries. Step out into the backyard, for example, push your thumb and index finger into the root zone of a patch of grass, and bring up a pinch of earth. You will likely be holding close to one billion individual living organisms, perhaps ten thousand distinct species of microbes, most of them not yet named, cataloged, or understood. Interwoven with the thousands of wispy root hairs of the grass would be coils of microscopic, gossamer-like threads of fungal hyphae, the total length of which would best be measured in miles, not inches. That's in just a pinch of earth. In a handful of typical healthy soil there are more creatures than there are humans on the entire planet, and hundreds of miles of fungal threads.

Some zealous soil ecologists have recruited small armies of graduate students and marched them into forests and grasslands to compile a complete inventory of subterranean life. Within a dimension of one square yard (about one square meter), they typically uncover billions of microscopic roundworms called nematodes, anywhere from a dozen to several hundred of the much larger earthworms, and 100,000 to 500,000 insects and other arthropods (species with hard exoskeletons). And that's in addition to the astronomical numbers of fungi, single-celled bacteria and protozoa, and other creatures that don't fall into these major groups. Many of the arthropods are very tiny and can be spotted only with a magnifying lens. Some defy classification; they simply have never been seen before. Even in well-studied areas, new arthropods and other multicellular species of unknown function are routinely encountered.

The numbers are staggering, the biodiversity fascinating, and the potential for discovery unsurpassed by any other habitat on Earth. Yet we have spent more time and effort examining small patches on the surface of the moon and Mars than exploring the subterranean habitat of our own planet. The words of Leonardo da Vinci ring as true today as they did five hundred years ago: "We know more about the movement of celestial bodies than about the soil underfoot." Even in modern laboratories, scientists are lucky if they can come up with the right nutrient mix to culture and study *1 percent* of the microbes found in a typical soil sample. This poor success rate is due in part to the complex interdependence between subterranean organisms. They can't survive when isolated from their neighbors. Until very recently, we knew next to nothing about the 99 percent of soil microbes that we could not raise in captivity except what their rotting cellular remains looked like under a microscope.

NEW TOOLS, HOWEVER, have opened a window to the underground, ushering in a cornucopia of subterranean discovery. Many of these tools have been borrowed from the tool chests of molecular biologists. The same molecular techniques that allow modern crime labs to detect a fragment of the genetic material left by a suspect at a crime scene allow modern soil labs to probe for evidence of specific organisms or characterize the full range of microbial diversity in a pinch of soil or a rock sample from the deep Earth. This still leaves many questions unanswered, such as the function of the many genetic types found. However, by identifying similarities in the genetic codes of newly discovered organisms and those we are already familiar with, scientists are often able to determine the ecological role played by previously unknown genetic types. At this point, the scientific community is grateful simply to have an opportunity to begin to quantify and catalog all that we have been missing.

In addition to the revolutionary breakthroughs in molecular biology, advances in engineering are providing scientists with new tools for reaching the remote habitats of creatures of the underground. New specialized drilling equipment and sterile techniques have been designed to tunnel into microbial habitats more than 10,000 feet (3,000 meters) below the rocky continental crust and seafloor, and retrieve samples that are completely free of microbial contamination from the surface and soils above. Such explorations have verified the existence of independent ecosystems that thrive in these deep subsurface habitats without sunlight, oxygen, or traditional carbon food sources, and at temperatures often exceeding the boiling point of water. On another front, advances in fiber optics, camera miniaturization, and radio tracking devices have given scientists unprecedented opportunities to study the behavior of burrowing animals within their subterranean habitats. Ecologists have gained valuable information for protecting species at risk, such as the black-footed ferret. Such information can also be used to develop humane control procedures for burrowing animals that have become agricultural pests.

Every venture into the underground using these new technologies takes us into unexplored territory filled with unexpected delights. The comment of Alice as she wandered through Wonderland—"curiouser and curiouser"—comes to mind. The textbooks can't begin to catch up. Our old notions of biology are being turned topsy-turvy. We are beginning to realize what "surface chauvinists" we have been in our myopic vision of life on the planet, blind to all but the most obvious of subterranean creatures. The latest scientific data suggest that the total biomass of the life beneath our feet is much more vast than all that we observe aboveground. Despite the preponderance of new evidence, our overreliance on visual experience to define our sense of reality makes this notion almost impossible to accept. We see the density of a rain forest, or the enormity of a giant redwood tree, and can only shake our heads incredulously at the possibility of another living world, a hidden subterranean biosphere, more immense than the grand scale of life aboveground.

With each new subterranean discovery, it becomes more apparent that the niche occupied by *Homo sapiens* is more fragile and much less central than we once thought. Just as the astronomical discoveries of Copernicus forever changed our notion of our physical place in the universe, our new knowledge of the magnitude and genetic diversity of Earth's subterranean world will forever change how we think about our place in the evolutionary "tree of life." This revolution began inauspiciously in microbiology and soil ecology but has now spread to encompass the much broader disciplines of evolution and biology. And we have already benefited in practical ways from this revolution. For example, we are putting soil microbes to work for us to combat plant and human diseases, and to help in the cleanup of our toxic wastes.

THE STATUS OF SUBSURFACE biology today is reminiscent of where marine biology was fifty years ago, when Jacques Cousteau was first perfecting his Aqua-lung for exploration of another hidden realm—the oceans. It might sound odd, but my own experience with scuba diving was the spark that motivated me to write this book. Since taking up the sport just a few years ago, I have been awestruck by the scenic beauty and diversity of underwater life. I will always be grateful to the professional dive masters who have taken me and my friends on guided underwater tours. While we nervously stare at our gauges and try to avoid kicking each other with our fins, one of the guides will tap on his or her air tank and point—whoa! It's a giant tarpon here, a manta ray there, or some other spectacular pelagic swimming below. Then, with hand signals, the guide herds us over to a coral formation and has us hover until we see what he or she has spotted—perhaps some tiny sea creatures doing a ritual underwater mating dance.

In my professional life as an ecologist, I have had the rare pleasure of exploring and learning about a realm—the subterranean world—that is just as diverse and fascinating as the oceans, but even less well known. It has also been exciting to bear witness to a scientific revolution in progress. I don't think I exaggerate when I say that what I have discovered of the mysterious underworld of our planet not only adds tremendously to my enjoyment of the outdoors and appreciation for the full scope of Earth's biodiversity, but enhances the quality of my day-to-day life. As a university professor and teacher, my natural inclination is to want to share this.

Knowledge of the incredible beauty, diversity, and activity of the subterranean world completely alters one's perception of the landscape. Gazing out over a barren plain becomes an experience similar to that of gazing out at a wide expanse of turquoise sea. On my frequent walks along the trails of the Ellis Hollow woods near my home in upstate New York, I get the best of both worlds. There is, of course, the surface life to enjoy—the grassy meadows, the stands of maple, beech, and hemlock trees, the deer and scurrying rabbits being chased by my Labrador retriever companion. But I know this is just the icing on the cake. I am keenly aware of all that I cannot see—the thriving communities of unusual life forms beneath my feet, some of them perhaps several thousand feet below, playing their role in the cycling of the elements and other functions important to all life on the planet. Only hints of subterranean activities are revealed at the surface: The "earthy" perfume of soil bacteria known as actinomycetes, the small mounds of digested soil or "casts" left behind by earthworms, and the openings to the burrows of larger animals.

My doctoral training was in ecology and plant biology. My research program within an agriculture department at Cornell University for the past fifteen years has followed up, more or less, on that experience. Not surprisingly, my interest in soil organisms began with those that affect soil fertility and plant health, and in particular those that form mutually beneficial partnerships with plants. Professional interest evolved into a more personal interest as I was exposed, via scientific journal articles and discussions with colleagues in other disciplines, to the growing excitement generated by the revolutionary developments in many other aspects of soil biology. Writing this book has given me the opportunity to delve into those areas peripheral to my own research much more thoroughly and to discuss the details with scientists who have been at the center of the most exciting discoveries.

This book is by no means intended as a comprehensive treatment of the subject of soil ecology. My goals are modest: To introduce you to a few of the most intriguing creatures of the underground, and to the sometimes equally intriguing scientists and explorers who have studied them, from Charles Darwin and Lewis and Clark to those whose names may be less familiar. You may be surprised to learn of the many ways in which the diverse subsurface biosphere is relevant to our everyday lives and to the environmental issues we will be confronting in the twenty-first century. I hope to serve as a subterranean "dive guide" of sorts as we take a journey together into a mysterious world we are just beginning to comprehend.

As humans, we are a particularly "subterranean-impaired" bunch. We are oversized, solar- and oxygen-dependent, and genetically programmed to think two-dimensional surface-space. These characteristics tend to exclude us from entering or fully appreciating the most fascinating habitats of the underground. What an adventure it would be if we could become microscopic spelunkers for a day in order to search for new life forms in the dark damp caverns of our backyard soil. Imagine rappelling our way down through a spectacular array of clay crystal formations, the beams from our hard-hat headlamps crisscrossing wildly as we try to catch glimpses of bizarre creatures scurrying about between the particles of sand and silt as big as giant boulders. We would get another perspective entirely if we could hop inside a microscopic submersible to join the unusual aquatic creatures that swim the narrow water channels of the upper soil horizons. If we could dive deep enough to roam the cold dark aquifers, or deeper still into the steamy zones heated by magma upwellings from the Earth's mantle layer, even stranger worlds would be revealed. Such a "fantastic voyage" is impossible, of course, but with some imagination, there is nothing to stop us from taking an initial dip into the underground for an overview of what can be found there.

The subterranean is not one world but many. It is filled with many unique habitats, and the denizens of these habitats range in size from the microscopic bacteria to the easily visible earthworms and burrowing animals (figure I.1). Over evolutionary time, the activities of this diverse flora and fauna transformed the uppermost layer of the Earth from one of sterile, pulverized rocks and minerals to one that could sustain life both above- and below-ground.

When soil scientists visualize their subject of study they do not imagine a loose pile of dirt, any more than a doctor would imagine his or her patients as a pile of body parts. Soil scientists think in terms of soil "profiles," which define the complex organic whole—the pattern of horizons or layers of minerals, organic matter, and living organisms as one ventures down from the surface toward bedrock. Soil profiles are used like fingerprints to classify soils in terms of their physical and biological composition and the history of their formation. Soils are referred to by taxonomic names based on their profiles, just as biologists use the binomial genus-species classification system to identify organisms. For those who know the taxonomy, a soil profile's name immediately reveals what type of rock "parent" material

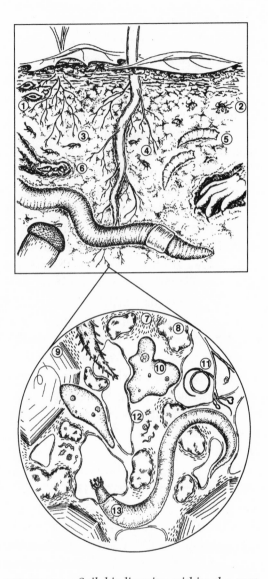

FIGURE I.1 *Soil biodiversity within the root zone. Depicted in the upper diagram are: (1) buried seeds, (2) water bears, (3) springtails, (4) mites, (5) insect larvae, and (6) ants. Depicted in the lower, magnified view are: (7) clay particles, (8) silt, (9) sand, (10) protozoa, (11) fungi, (12) bacteria, and (13) nematode. Illustration by Tamara Clark.*

the soil was formed from, details of its biological and climate history over geological time, the topography of the region, and its suitability for farming, skyscraper construction, or other uses.

I'll never forget my experience at a soil-judging contest when I was a graduate student at the University of California at Davis. I had accompanied a friend who was working toward his Ph.D. in soil science and was representing our school in a finals competition against other major land-grant universities. We could easily have been at a more conventional sporting event: The atmosphere was tense, and the contestants were focused, their adrenaline flowing. Soil pits about six feet deep had been dug in several locations in a meadow and up along a hillside to expose the profiles. Teams of students were in these pits with color charts, reference books, and meter sticks to measure the width of various horizons. The goal was to classify correctly the soils in each location. Winner take all. I was just an observer, so I listened while the Davis students debated in a fevered pitch among themselves. "Look, look at the cambic horizon; this has to be postglacial." "An Inceptisol!" one of them yelled out, pointing to one of the layers. "No way!" another grumbled. "See the shallow penetration of humus, and that argillic-like horizon? This is an Alfisol if I ever saw one." I realized then that this special language of the underground, and a fascination with profiles, played a role in attracting many to this field of study.

Although my soil scientist friends would cringe at the notion of a typical soil profile, for purposes of introduction I have attempted one in figure I.2. The upper "O" (for "Organic") horizon is not truly a soil by most people's definition because it lacks the mineral components of soil—sand, silt, and clay. Commonly referred to as the "litter layer," it may be composed of newly fallen plant debris, decaying logs, insect and animal carcasses at the top, and rich, partially decomposed organic matter and earthworm casts near the bottom. This layer can be several inches thick in a temperate forest or grassland, or almost nonexistent in a desert or tundra environment. The O horizon is the interface between the surface and subsurface realms, and many of the creatures one finds here inhabit both worlds. Familiar arthropods such as ants, sowbugs ("rolly-pollies" we called them as kids), millipedes, and beetles are among this group.

The larger burrowing animals also fall into the category of part-time soil dwellers. This group includes moles, ground squirrels,

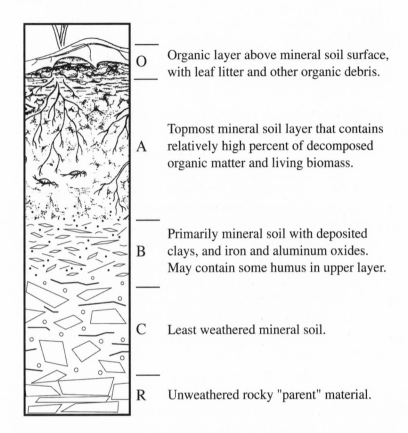

O — Organic layer above mineral soil surface, with leaf litter and other organic debris.

A — Topmost mineral soil layer that contains relatively high percent of decomposed organic matter and living biomass.

B — Primarily mineral soil with deposited clays, and iron and aluminum oxides. May contain some humus in upper layer.

C — Least weathered mineral soil.

R — Unweathered rocky "parent" material.

FIGURE I.2 *Typical soil profile showing the primary soil layers, or "horizons." Illustration by Tamara Clark.*

prairie dogs, ferrets, burrowing owls, rabbits, aardvarks, and foxes, to name a few. Many of them spend a lot of their time within the litter layer searching for food items such as fallen fruit, nuts, and tasty earthworms and insects. Others spend most of their time below-ground feeding on plant roots. Still others are predators near the top of the food chain that roam the surface or the tunnels below in search of small prey. The burrowing animals as a group are among the most intelligent creatures of the underground, and many live in communities with a complex social structure. Prairie dog colonies often contain thousands of members and spread over an area of several square miles or more.

Fortunately for soil ecologists, the vast majority of activity takes place within the litter layer and the top layer of actual soil just below it, called the A horizon. The A horizon is typically four to sixteen inches (ten to forty centimeters) deep. It often contains the highest densities of plant roots and soil organisms of the entire soil profile. It is also often rich in decomposed organic matter as well as a mixture of sand, silt, and clay minerals. Much farm labor is devoted to managing this "plow layer" zone which is so important to crop productivity.

We tend to think of plants, whether crops, wildflowers, or gigantic trees, primarily in terms of their aboveground foliage displays. This is only half the story, of course. Plants are in fact unique in that they equally and *simultaneously* inhabit both the surface and subsurface worlds. They are the great mediators between the two realms. Their leaves gather carbon and energy through the process of photosynthesis, while their roots drill for water and mine the soil for essential nutrients. Plants eventually are eaten or die and decompose, serving as the base of the food chain for both surface and subsurface ecosystems.

The soil zone immediately around plant roots is a unique, highly populated habitat because it is rich in sugars and other nutrients that leak out from living roots or become available as small root hairs die and decompose. The roots themselves account for much of soil biomass. If we could magically take an upside-down stroll through the underground, we would experience a dark "forest" of roots as dense as the aboveground display of tree trunks, branches, and leaves.

The roots of most plants do not work alone in their quest for water and nutrients. They are helped by beneficial types of soil fungi, collectively called mycorrhizal fungi, that attach to roots and often form a below-ground network between plants of different species. Not all fungi are so helpful; some cause plant diseases, while others parasitize neighboring soil organisms. But the majority of fungal species are harmless and live strictly off of decaying wood and other organic debris, playing an important role in decomposition and nutrient cycling. The threads of fungi of all types that crisscross throughout the O and A soil horizons provide an added benefit for soils by sewing small soil particles together into aggregates. This improves soil "tilth," or structure, improving drainage and aeration of the subsurface.

The thousands of species of bacteria found in the top two soil horizons play a vital role in the biogeochemical cycling of elements that is important to the sustainability of all life on the planet. Many break down organic matter to basic nutrients that can be absorbed by roots of plants or used by other soil organisms. There are very few waste products, pollutants, or toxins that cannot serve as food to one bacterial species or another. A few bacterial species found in the upper soil zones are pathogenic to plants, animals, or even humans. (Yes, Mother was right, you should wash your hands after playing in the dirt.) But others form important, mutually beneficial symbiotic relationships, such as the very important nitrogen-fixing bacteria that attach to roots and supply the plant with nitrogen in exchange for sugars.

There are many tiny micro-arthropods that inhabit the A horizon and the litter layer. Mites, which resemble tiny spiders, and springtails (some of which literally spring about when they come to the surface) are particularly abundant in most soils but are easily missed without at least a magnifying lens. As many as 100,000 springtails might be found in a square yard of forest soil. Springtails are also known as snow fleas because they sometimes appear in large numbers on melting snowbanks in early spring. Also very common are the non-arthropod, microscopic water bears. They were given this name by the famous nineteenth-century naturalist Thomas Huxley, who thought they resembled bears (see figure I.3). As their name implies, they are essentially aquatic and swim along the water films adhering to mosses and plant debris. Water bears are famous for their ability to dehydrate to a small percentage of their normal water content and hibernate until conditions become favorable again (for up to one hundred years if necessary!). As part of the soil food web, the water bears and arthropods feed on organic debris, nematodes, or fungi and other microbes that are abundant in the upper soil layers.

Earthworms, justifiably the best known and most beloved of soil creatures, can be found in the litter layer at night, when they creep up from deeper horizons to harvest fallen leaves. Walking through a deciduous forest, it is easy to spot the openings to earthworm burrows if one knows to look for the bouquets of dried leaf petioles sticking up from them. Often the first to attack freshly fallen plant debris, earthworms are extremely important as "biological blenders"

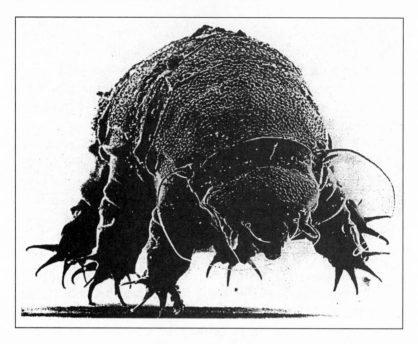

FIGURE 1.3 *Scanning micrograph of a water bear (tardigrade), magnified 150 times. Water bears usually inhabit damp soil or mosses and leaf litter. Courtesy of John H. Crowe, University of California at Davis.*

of the underground. Their burrows create biopores that serve as an oxygen pipeline for plant roots and other subsurface life. Earthworms are appropriately named because as they tunnel through the soil horizons they consume huge amounts of dirt, gleaning it for useful food items. Their waste droppings, or casts, are rich in organic matter and nutrients and are heavily colonized by small arthropods and microbes, who break the material down further. In many ecosystems, quite possibly including your own backyard, the uppermost "soil" is composed almost entirely of casts left behind by earthworm feeding.

Continuing our downward journey, the next layer is the B horizon. Roots and soil creatures are not difficult to find here, but their numbers are fewer. Arthropods are rare at this depth, but one would encounter earthworms (which can easily burrow to a depth of three feet, or one meter), fungi, nematodes, and bacteria. The B horizon is often high in clays, and some of the highly decomposed, black, gummy organic matter known as humus is sometimes near the top of the layer.

The C horizon, which typically begins three to six feet below the surface, is relatively devoid of organic matter. Soil life here is often exclusively microbial. The numbers of bacteria at this depth are usually less than one thousand within a gram of soil (a pinch), compared to populations approaching one billion per gram nearer the surface.

Until quite recently, scientists believed that the region below the C horizon was incapable of supporting life. But in just the past couple of decades, we have discovered thriving microbial ecosystems thousands of feet below our common soils. Some are so deep that they are heated by the radioactivity and upwellings of hot magma from Earth's interior. The creatures that inhabit these environments are called "extremophiles" because they are living without oxygen or light, and at temperatures that often exceed the boiling point of water.

THE MORE WE LEARN about the spectacular biodiversity of the underground, the more questions we come up with. How and where did it all begin? How did surface and subsurface creatures co-evolve such an efficient recycling system, one that has sustained life on Earth for three and a half billion years?

Addressing questions such as these will be part of our journey, and will require some digging into the past. We begin our story in chapter 1 with the very origin of the Earth and its soils, and the evidence that life on our planet may have first emerged in a subterranean environment. In chapters 2 and 3, we examine the Earth's most ancient life forms, the extremophiles, and learn how the discovery of their amazing genetic diversity has led to a complete overhaul in our thinking about the course of evolution and our place within the tree of life. In chapters 4 through 6, we learn about a few inconspicuous creatures of the underground (including a bacterium, a fungus, and the earthworm) and their tremendous impact on the cycling of elements and the flow of energy essential to all life on our fragile planet. Chapter 7 discusses the dual nature of soils with regard to deadly plant and human diseases: A few soil microbes have caused much human suffering, while others provide us with some of our most powerful antibiotics. Chapter 8 describes the tragic history of human interactions with prairie dogs and other burrowing animals. The final chapter explores the impact of human activities on the soil resources important to our food security, and the potential for using soil microbes for bioremediation of damaged soils.

No need to don special gear, other than perhaps reading glasses, for this journey. Just find a quiet spot, a comfortable chair, and let's dive in.

Part One

ANCIENT LIFE

1

ORIGINS

*The origin of life appears. . . to be almost a
miracle, so many are the conditions which would
have had to be satisfied to get it going.*

—FRANCIS CRICK, *LIFE ITSELF: ITS NATURE AND ORIGIN*
(1981)

*Why, who makes much of a miracle?
As to me I know of nothing else but miracles,. . .
Every cubic inch of space is a miracle,
Every square yard of the surface of the earth is
spread with the same,
Every foot of the interior swarms with the same.*

—WALT WHITMAN, *LEAVES OF GRASS* (1855)

THE EARTH WAS NOT CONSTRUCTED WITH A DELICATE HAND.
It was hammered into shape slowly, by the brute force of a meteor
bombardment that lasted hundreds of millions of years. The soils,
the seas, and our primitive microbial ancestors emerged in the midst
of apparent chaos and catastrophe. The process began billions of
years ago as our entire solar system was congealing from a swirling
cloud of hot gases and nuclear ashes left behind by exploded stars.
Some of the objects colliding with the Earth at this time were planetesimals—objects as big as small planets. The kinetic energy released
by these impacts literally shook the Earth to its core and melted much
of the rocky crust and interior. Some chunks of the planetesimals and
meteors became permanently embedded in the Earth, while other
pieces were sent hurtling off into space like giant shrapnel. The mass
of the primordial Earth accumulated slowly, like a globe that grows

as a sculptor slaps on clay, one handful at a time. With greater size, Earth increased in its gravitational force, attracting even more of the wandering debris of space.

It is hard to come up with a specific date of birth for our planet, given its gradual development. Basing their calculations on the "radioactive clock"—measurements of the level of radioactive decay of certain elements found within the Earth's crust, such as uranium and lead—most geologists place the Earth's age at about four and a half billion years. The Earth went through horrendous growing pains during its first billion years. Just as the frequency of meteor impacts began to decline, violent volcanic eruptions began to spring up around the globe as the planet's hot interior "degassed." When the Earth's surface temperature finally began to cool, the massive volume of water vapor in the atmosphere condensed and poured down from the heavens in fierce rainstorms of truly biblical proportions. The torrential rains lasted millions of years, creating our oceans—the hydrosphere as we know it—in the process.

The original igneous and metamorphic rocks on the Earth's surface, left behind by volcanic eruptions and upliftings from the mantle layer below, were washed by the relentless rains, and their minerals flowed into the oceans. This was an essential first step in the formation of primitive soils that would eventually support a vibrant plant and animal life. These primitive soils lacked organic matter but contained sand, silt, and clay minerals in various proportions.

Clays are unique among the mineral components of soil. They are chemically reactive, microscopic, crystal-like structures that form out of saturated solutions of silicate and metal oxides. Sand and silt, in contrast, are large, chemically inert particles formed by the simple weathering and pulverization of rock. Some clays are crystallized deep within the Earth's mantle layer, at high temperature and pressure, and then brought to the surface by the churning motions of the Earth. This process is driven by radioactive heating deep within Earth's mantle and is part of the same plate tectonic geological cycle that gradually moves the continental crusts.

HOW THE STARDUST components of our planet managed to buck the thermodynamic tendency for disorder, and organize into the intricate design of living systems, is a puzzle that has perplexed scientists for decades. This much is known: Life—the biosphere—originated sometime within those tumultuous first billion years of Earth's history. Microfossils discovered in recent years, and larger fossils formed by visible colonies of bacteria that clump together in mats called stromatolites, provide unequivocal evidence that microbial life was present as early as three and a half billion years ago, perhaps even earlier. Given what we recently have learned about the setting just prior to the emergence of these creatures—meteor bombardment, an epidemic of volcanic eruptions, and intense ultraviolet (UV) radiation (there was no ozone filter in the upper atmosphere)—many scientists are becoming convinced that the ancestors of Earth's first life forms must have originated well below the surface. Any new species that might have ventured out from Mother Earth's protective womb in those early years would have been quickly destroyed by one surface catastrophe or another, its evolutionary path nipped in the bud. The young Earth was like a war zone where the safest place to be—the *only* place to be—was underground.

The notion of the underground as the cradle of life is contrary to a popular theory held throughout much of the twentieth century that life began in a shallow body of water, or perhaps in the surface waters of the ocean, where evaporation might have concentrated just the right "primordial soup" of ingredients for life to emerge. This theory arose from Charles Darwin's speculation that life originated in "some warm little pond." Darwin wrote this in 1871 in an informal, private letter to his botanist colleague Joseph Dalton Hooker. It was not an idea that he had particular confidence in or intended to promote. Nevertheless, his followers took the remark quite seriously. The letter has been cited in virtually every book and review article on the subject of the origin of life since Darwin's day.

Darwin would probably be both surprised and a little dismayed to learn how much his casual comment influenced thinking on this matter in the twentieth century. In other writings, he made it abundantly clear that he felt the issue was best left to future generations, who would undoubtedly have a better foundation for tackling the subject. For example, in an 1881 letter to Nathaniel Wallich, curator of the Calcutta Botanical Gardens, Darwin refers to the issue as *ultra vires* (beyond the powers) of science at the time: "You expressed quite correctly my views where you said that I had intentionally left the question of the Origin of Life uncanvassed as being altogether *ultra vires* in the present state of knowledge."

Although it is possible that the details of the origin of life may forever be *ultra vires*, we have many exciting new leads to follow as we enter the twenty-first century. Most of these point toward a subterranean environment rather than a "warm little pond" as the cradle of life—possibly within the murky sediments of the ocean floor or deep within the water-filled pore spaces of the continental crusts. As we shall see, support for this idea goes well beyond the fact that the underground would have been the safest refuge from the violence and climatic turmoil of Earth's first billion years. The underground was also the place where the essential ingredients for primitive biochemistry were to be found, and where today we find bizarre microbes believed to be the direct descendants of Earth's first life forms.

JUST A COUPLE OF YEARS before Darwin wrote his frequently cited letter speculating about the warm little pond, another famous naturalist of the day, Thomas Huxley, had published a bold and widely read essay entitled "On the Physical Basis of Life." Although Huxley agreed with Darwin that it was premature to attempt to pinpoint the origin of life, he explained that living organisms are constructed from atoms and that life's activities are ruled by the laws of physics and chemistry. Huxley reached very deep and stretched very far considering it would be another century before the field of molecular biology emerged. He was accused of religious heresy in many quarters, but this was nothing new for Huxley, who was already well known as a fearless and eloquent supporter of Darwin's evolutionary theory.

Huxley identified four elements as primary ingredients for the evolution of life—hydrogen, carbon, oxygen, and nitrogen. Modern chemical analyses verify that of the more than one hundred elements in our periodic table, these four account for more than 95 percent of the atoms found in the human body. The same is true for bacteria, fungi, earthworms, great white sharks, giant redwoods, you name it. This similarity in the elemental composition of all life forms (that we know of) is a point that Huxley also emphasized.

What is even more remarkable, however, is the similarity between the elemental composition of living organisms and that of the universe as a whole. Recent spectroscopy measurements of stars and interstellar dust confirm that the same four elements identified by Huxley as the main components of most of the biosphere also happen to rank within the top five in cosmic abundance. The miracle of life, as we shall see, lies in its complexity, not in the scarcity of start-up ingredients.

Hydrogen makes up more than 90 percent of all the matter in the universe, and more than 60 percent of the atoms in the human body. All of this hydrogen was formed in the fiery explosion of the "big bang" fifteen billion years ago. It is the simplest of all atoms, with a nucleus containing one proton and one neutron, orbited by a single electron. All of the other elements in the human body were forged some time later by the nuclear fusion reactions of burning stars. In these nuclear fusions, the nuclei of simple light elements, beginning with hydrogen, collide to form the larger nuclei of the heavier elements. As William Fowler said in accepting the Nobel Prize for his work on the origin of the elements in 1983: "All of us are truly and literally a little bit of stardust."

The Earth is by no means unique in the universe in containing the basic elements of life. In fact, due to the way things sorted themselves out in the initial formation of our solar system, the Earth has relatively less hydrogen, carbon, oxygen, and nitrogen than some of our planetary neighbors more distant from the sun. Nevertheless, the fact that we and all our biotic co-inhabitants are here is proof that the Earth contains enough of life's essential elements to build a thriving biosphere, *provided that a system for recycling those elements is in place.* Soil organisms play a central role in this recycling system, as we discuss later. The key question here is, why did life originate from these basic elements on our planet and presumably not on others?

Earth's great advantage as a life-generating planet lay not in a superior abundance of essential elements, but in the fact that many of these elements were combined into specific molecules that facilitated an evolution from geochemistry to biochemistry. The 1871 essay by Thomas Huxley goes on to identify three simple molecules that were essential for the formation of life on Earth: water (hydrogen and oxygen), carbonic acid (carbon, hydrogen, and oxygen), and ammonia (nitrogen and hydrogen). Huxley's assertion has stood the test of time. All modern theories of the origin of life recognize the important role played by these three molecules, and all concur that an abundance of one of them—water—is what is most unique about our blue planet.

It is within the milieu of water that the chemistry of all life as we know it takes place. Many of the other compounds essential for life are useful only in the presence of water. Their role is determined by whether they dissolve in water and by the effect of water on their electrochemical properties. Water is found nearly everywhere on Earth, both above and below the surface. Even in desert environments that appear dry and lifeless to us, thriving microscopic communities of subterranean organisms are often swimming happily within the thin water films that adhere to clays and porous rocks.

Prior to the appearance of life on our planet, it was within water and water-saturated soils and sediments that many important organic compounds (which contain both carbon and hydrogen) were first synthesized. Organic compounds such as amino acids, nucleotides, and lipids were the necessary building blocks for Earth's first proteins, genes, and cell membranes, respectively. The synthesis of these building blocks could have occurred spontaneously only if the basic thermodynamic laws of nature favored the chemical reactions, or if energy was supplied to overcome the thermodynamic barrier. Just as a ball tends to run downhill rather than uphill unless we supply energy to push it up, a chemical reaction tends to go in the thermodynamically downhill direction unless energy is supplied to make it do otherwise. Christian de Duve, the Nobel Prize–winning biochemist, felt that "the pathway to life must have been downhill all the way."

But life as we know it today does not just happen spontaneously. Biomolecules and their interactions with each other are highly organized and therefore go against the basic thermodynamic tendency toward entropy, or disorder. Living organisms have successfully battled entropy by evolving mechanisms to harvest external energy (the gathering of solar energy by photosynthetic organisms, for example) and use it to go uphill, to drive thermodynamically unfavored reactions. When an organism dies, this capacity is lost, entropy wins the day, and catabolic reactions break down complex biomolecules into their elemental components. The question is, how did those basic building blocks of life whose synthesis is *not* thermodynamically favored ever come into being before there were living organisms to harness the necessary energy?

One possibility is that some of the energy-requiring reactions occurred in the vicinity of, and were tightly coupled to, other reactions that were favored and released energy. Another possibility would have been fortuitous unpredictable energy input from external sources such as ultraviolet radiation or lightning strikes. In 1951, Stanley Miller, who was then a graduate student working under the direction of the physical chemist Harold Urey at the University of Chicago, conducted a landmark origin-of-life laboratory experiment that demonstrated this latter possibility. Within an apparatus of flasks, glass tubing, and condensing columns, he created an artificial atmosphere composed of hydrogen, ammonia, and methane and a water "ocean." He passed electric sparks through the gas to simulate lightning supplying energy for chemical reactions. Within about a week of this treatment, the liquid contents had turned a vivid blood-red color, and Miller decided it was time to make an analysis of the results. To the surprise of the researchers and the rest of the scientific community, 15 percent of the original carbon of the methane gas turned up in several different water-soluble amino acids. Although amino acids are by no means huge macromolecules like some of the proteins that they are part of, their synthesis from a mixture of such simple ingredients was a bit of a shock.

Among the other important products of the Miller-Urey experiment were formaldehyde and hydrogen cyanide. Formaldehyde was an exciting result because it has the capacity to self-assemble into a ringed sugar molecule known as ribose, which is an important component of our genetic material—RNA (*ribose* nucleic acid) and DNA (deoxy*ribose* nucleic acid). Hydrogen cyanide was initially considered an undesirable and toxic by-product, but later it was shown that it could lead to the production of adenine, another molecule of tremendous importance in biochemistry. Adenine, like ribose, is a component of RNA and DNA as well as a part of the structure of adenosine triphosphate (ATP), life's most important molecule for the storage and transfer of chemical energy.

By the 1970s, some scientists were beginning to conclude that the puzzle of the origin of life was all but solved. Cooking up the "stuff of life"—amino acids, nucleotides, lipids—was proving to be a piece of cake. Add a dash of energy, perhaps lightning or UV radiation, to some common chemicals and . . . *voilà!* It wasn't life exactly, but the results were interesting enough to stir the imagination.

Spirits were substantially dampened in the 1980s, however, when new evidence collected by geologists and astronomers began pouring in to indicate that the assumptions about Earth's primitive atmosphere and oceans in the Miller-Urey and subsequent experiments were off the mark. Hydrogen gas is an element too light to have ever been held in our atmosphere in any significant quantities by Earth's gravitational field. Also, while methane and ammonia may have been present in some phases of the evolution of Earth's atmosphere, the concentrations assumed by Miller and Urey were almost certainly too high.

Just when it was beginning to appear that the brilliant experiments of Miller-Urey and others had been for nought, discoveries within some of Earth's most obscure subsurface environments renewed hopes. It turns out that hydrogen, methane, ammonia, and other chemicals used by the pioneer origin-of-life researchers for synthesizing the basic building blocks occur in abundance near deep-sea hydrothermal vents and in certain regions deep within the Earth's crust. These subsurface environments are hot—often near the boiling point of water—because of heat generated by radioactivity, compression, and upwellings from the mantle layer below. Energy budget calculations suggest that this thermal energy could be used to drive the synthesis of amino acids and nucleotides.

Although this discovery offered a reasonable explanation of how life's basic building blocks might have been synthesized, as the twentieth century ended we were still very far from resolving the mystery of life's origin. Ironically, it was one of the century's major achievements in biology—the discovery of DNA's function and complex molecular structure—that became the core of the impasse for origin-of-life researchers.

Biologists marvel at the size and complexity of DNA: Two nucleotide chains intertwined in the famous double-helix shape, with the precise sequence of the thousands of nucleotides in each chain determining the genetic code. Even with an abundance of nucleotides, how could these elegant genes of ours have been put together? And the same question can be asked with regard to the formation of complex proteins from a soup of amino acids. Expecting such things to happen simply by supplying energy is too far-fetched. As the Australian physicist Paul Davies put it, that would be like "exploding a stick of dynamite under a pile of bricks and expecting it to form a house."

In modern cells, individual strands of DNA form the template for their mirror-image replication during cell division. This process is catalyzed by protein-enzymes that are very large macromolecules themselves, often comprising hundreds of amino acids folded up in peculiar ways. The various folds and prosthetic-like extensions of the enzyme help to hold the correct nucleotides close together and in just the correct orientation during the DNA synthesis process, thus reducing the energy requirement. But in a prebiotic world, before there were living cells with DNA instructions for the sequencing of amino acids, how would such enzymes have been constructed? It's a mind-boggling catch-22 problem of interdependence. It is nearly impossible to imagine the synthesis of DNA (or RNA for that matter) without a protein-based enzyme to catalyze the hooking together of nucleotides, and it is equally hard to imagine how these required enzymes, composed of hundreds or thousands of amino acids, could be produced without instructions from DNA and single-stranded RNA "working copies" of the DNA code.

AS WE ENTER THE TWENTY-FIRST CENTURY, many scientists feel that the puzzle of the origin of life has been narrowed down to two fundamental questions. The first is a "chicken or the egg" kind of problem: Which came first, genes carrying the code necessary to create proteins, or protein-enzymes that could catalyze the synthesis of genes? The second question is, whichever one came first, how was it created? There are numerous theories attempting to resolve these issues, and one of them casts the soil itself—clay minerals to be more precise—in the leading role.

The idea is not new. For thousands of years, before any scientific explanation was available, humans have linked soil and life. In the second chapter of the Book of Genesis, we are told: "The Lord God formed man of the dust of the ground." Appropriately, in this Bible story the name of the first person to roam the Garden of Eden was Adam, a name based on the Hebrew word for soil or clay, *adama*. And the Latin name for man, *homo*, used in our genus-species designation *Homo sapiens*, is derived from *humus*, which can be translated as "of the soil or earth." Many Native American cultures feel a particularly strong connection to the soil, as reflected in Chief Seattle's response in 1852 to the U.S. government's offer to purchase his tribe's Northwest territory: "How can we buy or sell the sky or the land? The idea is strange to us. . . . We are part of the earth and it is part of us."

The scientific explanation for a connection between soil and life is founded on basic chemistry. It suggests that the electrostatically charged surfaces of clay minerals served as primitive enzymes and provided the catalytic sites of Earth's first complex biosynthesis. Some of the macromolecules that resulted may have been simple chains of nucleic acids, while others were chains of amino acids. Another, more radical, version of this theory holds that clay crystals were themselves the Earth's first self-replicating "low-tech genes." These growing clay crystals were not really alive, but they were able to evolve and to shape their environment in simple ways. The survival and replication of some of these crystals might have been favored by the incorporation of amino or nucleic acids into their structure. Through evolutionary time, as the organic macromolecules grew in complexity and efficiency, they took over replication and synthesis functions, and the clay infrastructure was discarded.

Clays, when viewed with the naked eye, are hard to distinguish from the other components of "dirt" and admittedly do not seem very special. However, when viewed through an electron microscope, their crystalline structure and exquisite beauty become apparent (figure 1.1). At the molecular level, they form geometric patterns that comprise oxygen and silicon (or other metal) atoms joined together into ribbons or nets. These ribbons or nets stack together like plates and are held together by electrochemical forces. When wetted, the water molecules lubricate the area between the layers and allow the plates to slide easily across each other. This gives clays their "plastic" property, which makes them easy for artisans to mold into bowls, vases, and other objects of all shapes and sizes. Most children learn to appreciate this property at an early age in "mud pie"–making marathons with their friends. Even as adults, some of us can get a squeamish kind of thrill by wading into a lake and squeezing the slippery clay sediments of the bottom with our toes.

The microscopic size and layered, porous structure of clays give them some unique characteristics. They have an incredibly high surface-to-weight ratio. A single gram—a pinch—of clay powder can

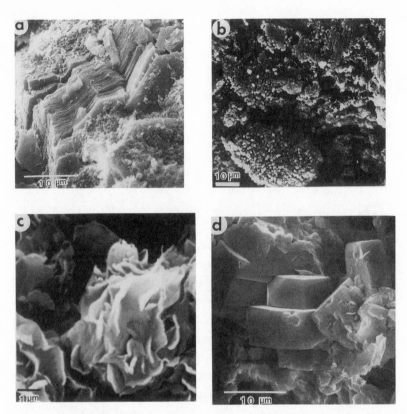

FIGURE 1.1   *Scanning electron micrograph of various clay crystals. From A. L. Senkayi et al.,* Clays and Clay Minerals *32 (1984): 259–71.*

have the surface area of a baseball diamond! The other mineral components of soil, sand and silt, have a much larger particle size than clays but less exposed surface area. To illustrate the distinction in size and weight, consider that a particle of sand dropped into a glass of water falls about one inch in one second, while a particle of fine clay remains suspended and can take two hundred years to fall the same distance.

Because the exposed outer atoms of clay surfaces are electrostatically charged, atoms or molecules of opposite charge are attracted to them. This interaction can change the characteristics of the clays themselves and affect the chemistry of the surrounding medium. Farmers and gardeners know that the clays in soils are important for crop nutrition because they hold certain essential nutrients, such as positively charged forms of calcium and magnesium, to their surfaces and prevent them from being washed below the root zone by rains or irrigation water. Potters know that certain clays fire into a deep red color in the kiln as long as plenty of oxygen is made available. Such clays contain iron atoms that chemically react and combine with the oxygen, causing the rich red color. The iron-containing hemoglobin of our blood behaves in a similar way, turning a bright red when oxygenated.

Among the types of molecules that can be attracted to and held by the charged surfaces of clays are many organic compounds, including some forms of amino acids and nucleotides. The hypothesis that clays served as the template for the sequencing of simple proteins or genes is based on this fact. In addition to their charged surface, clay crystals have a very convoluted shape with many nooks and crannies, which could serve to bring pairs of amino acids or nucleotides of particular sizes or shapes into the correct orientation to facilitate synthesis. This is similar to the mechanism by which the large folded protein-enzymes of modern cells catalyze the biosynthesis of macromolecules.

The speculation that clays might have served this catalytic role in the origin of life is half a century old, having been first advocated by the British physical chemist John Desmond Bernal in the early 1950s. Experiments in the 1970s demonstrated that a common type of clay known as montmorillonite (named for the French town of Montmorillon where it was first quarried) could catalyze the sequencing of specially prepared amino acids. Some of the protein-like molecular chains synthesized in this way were up to sixty amino acids long. In the 1980s, James Lawless, of NASA's Ames Research Center in California, and others found that nucleotides could be made to bind to various clays when certain metals, such as zinc or copper, were also present in minute quantities in solution. Then, during the 1990s, J. P. Ferris and his colleagues managed to attach nucleotides together into long chains using montmorillonite clays as catalysts. The procedure required some preparatory treatment of the nucleotides, but the results nonetheless provided support for the idea that clays could have catalyzed the first simple RNA genes.

The capacity to store chemical energy is an important requirement for life, and some clays have this capacity. Montmorillonite, as well as other common clays such as kaolinite and illite, are chemically reactive with ATP, the molecule that serves as the "energy currency" in almost all living organisms. These clays affect the ATP's phosphate chemical bonds, which are crucial to the energy transfer properties of this important biomolecule. Lelia Coyne, of San Jose State University, and others have found that kaolinite clays can, on their own, store energy (gathered from radioactive sources), then release it when disturbed by environmental factors such as wetting and drying or temperature fluctuations.

Graham Cairns-Smith, a prominent chemist at the University of Glasgow, has been the strongest proponent of an even more important role for clays in the origin of life. He suggests that clays not only served as crude enzyme catalysts but were the precursors of today's genes. This theory is based on the fact that the unique sequence of atoms on a clay surface forms the template for the synthesis of the mirror-image replication of those atoms, similar to the way in which the nucleotide sequences of DNA and RNA strands are the template for their replication. Also, as clay crystals self-replicate, their structure often takes on irregularities (mutations, if you will), and these irregularities become repeated (figure 1.2). Clay crystals are "aperiodic": although they have an organized structure, the patterns are more like those in an exotic hand-woven tapestry than a factory-made wallpaper. The pattern irregularities in clays are "heritable" just as the mutations of real genes are: They get passed on to layers or segments of clay crystals that break off and become the template for new crystals.

The famous quantum physicist Erwin Schrödinger speculated on the molecular nature of our genes in the 1940s, before their structure had been discovered. He predicted that they would be aperiodic crystals, because these could hold replicable "information" based on their unique sequence patterns. He also pointed out that such crystals would allow for evolution by occasional substitutions of particular links in their surface chain of atoms. As it turns out, the double-helix DNA unraveled by Francis Crick and James Watson several years later is a type of aperiodic crystal. And so are clays.

Part of Cairns-Smith's argument is that we will never uncover the secrets of life's origin if we limit ourselves to considering only the materials we see in living organisms today. "Search in vain for one wood abacus bead in a pocket calculator," he quips. The first organisms would have been very "low-tech" and made from different mate-

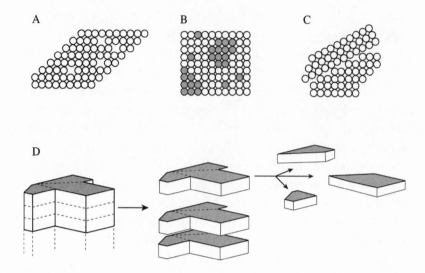

FIGURE 1.2   *Common clay crystal defects include (a) missing molecular components within a lattice; (b) substitutions of individual components; and (c) a misalignment between sections of a lattice. Defects such as these are analogous to genetic mutations and can be passed on to the "next generation," as segments or layers of clay crystals become detached, and (d) form the template for their own mirror-image replication. Illustration by Tamara Clark, adapted from © 1985 George V. Kelvin (Scientific American).*

rials than those that make up today's cells. Evolution would have started with the simplest and easiest approach. Clays, he argues, are likely precursors to genes because they spontaneously self-assemble (with the help of geological and hydrological forces), they self-replicate, and they would have been abundant on early Earth.

The idea of "clay genes" was considered a rather wild idea when first proposed by Cairns-Smith in the 1960s. It is still controversial, but supportive experimental data have since trickled in. The most crucial research has focused on how our nucleic acid–based genes could have evolved from a clay crystal beginning. It has been known for some time that amino and nucleic acids are sometimes found integrated into the structure of clays in nature. Some argue that these could represent a hybrid stage in the evolution from clay to organic genes. In addition to such circumstantial evidence, researchers have found that the crystallization process of some clay types is facilitated by, or may require, the presence of specific amino acids. Besides their effect on clay formation, amino acids and other organic molecules can exert powerful effects on the shapes and sizes of clay crystals by inhibiting the growth of certain faces.

One can imagine scenarios during evolution when the survival, competition, and self-replication of those clays containing organic molecules would have been favored. For example, a "defect" in clay shape caused by an organic molecule in the structure might have caused those crystals to be better able to cling to rock pores and not be washed away, or to stick to a surface that was constantly bathed in the essential compounds (nutrients) needed to replicate more of itself. Other clays may have replicated faster or better because of how their organic components affected the chemical interactions with the surroundings or the cementing between plates. Some clays may even have evolved the capacity to manufacture organic molecules for their own selective advantage.

The (yet to be proven) "genetic takeover" hypothesis begins with a simple RNA-like single-stranded nucleotide chain becoming embedded in a clay crystal and facilitating in some way the replication of the clay. In the initial stages, the crude RNA has a very minor, optional role, but through millions of replications, the RNA becomes more sophisticated. Because a complex organic macromolecule like RNA can hold more information than a clay crystal and achieve more intricate, selective control of operations, it begins to dominate the orchestration of the replication and catalytic functions of the clay. Eventually there is a complete "genetic takeover" by the RNA, and the clay matrix diminishes in size and importance until it completely disappears. Any clay crystals that might have taken in a nucleic acid for their own benefit would have sown the seeds of their own destruction.

The "clay gene" theory suggests that the progression toward life on our planet has in some ways mirrored the way a bridge is constructed. As a non-engineer, I marvel every time I cross a relatively simple steel bridge near my home. It seems there must have been some physically impossible step near the beginning when workers and steel beams floated on air to get the job done. The reality is that, at the beginning, special scaffold structures were undoubtedly built to support workers and equipment. After the bridge was completed and the scaffolds were no longer needed, they were removed. The lesson here is that construction, like evolution, almost always involves subtraction as well as addition. Cairns-Smith's hypothesis holds that clay crystals were the scaffold in the evolution of the first true genes.

During the past couple of decades, scientists have begun examining other common inorganic soil substances besides clays for "life-like" properties. The German chemist Günter Wächtershäuser has proposed a very comprehensive theory involving pyrite, commonly known as "fool's gold," that has attracted many followers. Pyrite is a very simple crystalline mineral that can be synthesized from ferrous iron and hydrogen sulfide, two substances that are abundant in many soil environments and are also found near deep-sea hydrothermal vents. The synthesis of pyrite is thermodynamically favored under many conditions and therefore *releases* rather than requires energy. This release of energy can be used to drive the synthesis of organic molecules, which then become bonded to the highly reactive surface of the pyrite crystals. Alternatively, the energy can be used to absorb carbon from carbon monoxide or carbon dioxide in its environment, similar to the "fixation" of carbon by photosynthetic organisms today. The surface chemistry of pyrite makes it superior to many clays in bonding to nucleotides. If experimental data can be obtained to back up the speculations regarding pyrite's powers, we may conclude that this "fool's gold" is much more valuable than real gold. Without it, there may never have been life on the planet, let alone a human civilization with a lust for precious metals.

THE DETAILS OF *HOW* LIFE may have originated continue to be elusive and the theories highly controversial. With regard to *where* life began, however, we seem closer to an answer. Recent breakthroughs in molecular biology and genetics tend to support the latest research in physical and soil chemistry—that life originated in extreme environments, perhaps the deep subsurface or within the sediments of deep-sea hydrothermal vents. New tools have allowed scientists to trace our genetic roots to our most primitive ancestors. We can examine the nucleotide sequences of the RNA of organisms living today and determine with considerable precision not only who is related to whom, but also who has the deepest roots in the evolutionary tree. It turns out that the living representatives of our most primitive ancestors are a diverse group of unusual microbes that inhabit some of the most hostile subsurface environments on our planet today. These "extremophiles," as they are called, are a fascinating story unto themselves, and the focus of the next chapter. The key point here is that because they represent our deepest genetic roots, a reasonable guess as to where life originated would be the subsurface habitats similar to those in which we find these unusual microbes today.

2

THE HABITABLE ZONE

*Heaven is under our feet as well as over
our heads.*

—HENRY DAVID THOREAU, *WALDEN* (1854)

As THE MIND OF THE HUMAN SPECIES EVOLVED, at some point
we began to contemplate the possibility of life beyond our planet.
Perhaps it was a starry evening, thousands of years ago, when some
primitive *Homo sapiens* stepped outside of his or her cave, gazed sky-
ward, and was the first to ask that existential question: Are we alone?
It has haunted us ever since. No one could have guessed, until just a
few years ago, that an important clue as to where to search for extraterrestrial life might be right here on Earth, beneath our feet, in the
"deep, hot biosphere."

In our recent search for the origin of life on Earth, we have made a
series of fascinating discoveries of microbial communities that thrive
thousands of feet beneath the surface, at extremely high temperatures and pressures, and without oxygen or light. Within the rocks
and clays, these "extremophiles" have access to water but often little
else that we would consider necessities. Many have been cut off from
sunlight for hundreds of millions of years, yet they eke out a living in
buried oil reserves or other organic carbon sources. Some gather carbon directly from carbon dioxide gas, and their energy is derived, not
from the sun or the consumption of ancient buried plant life, but
from hydrogen gas or inorganic chemicals found within the rocky
substrate. These amazing creatures are "primary producers" within
their dark realm, supporting entire ecosystems as the base of a food
chain, just as plants and other photosynthetic organisms do on the
surface. The apparent independence of these underground commu-

nities has completely revolutionized our thinking about life on our planet and elsewhere. It flies in the face of the lesson we all learned in high school biology—that all life is ultimately dependent on solar energy. Some scientists now believe that the microbial organisms at the base of the "dark food chain," with their bizarre metabolisms, may be the direct descendants of Earth's first life forms.

Astronomers and geologists concur that many of the solid planetary bodies in the universe are likely to have subsurface environments very similar to Earth's. The temperature and pressure conditions within the interiors of some of these planets could even maintain water in the liquid state. Since there are organisms that find the extreme conditions of deep Earth hospitable, why not the deep subsurface of Mars? Or Jupiter's moon, Europa? And if, as some suspect, life originated within Earth's subterranean, couldn't life also have arisen in one of the many similar environments elsewhere in the solar system, or in the wider universe? In our myopic view that only solar-powered life is sustainable, we have presumed that the "habitable zone" around any sunlike star had to be a distance of about one and a half Earth orbits, the zone in which we would expect *surface* conditions similar to ours. It now appears that the habitable zone, within our own planet and throughout the universe, has been substantially underestimated.

FOR THE EXPLORERS OF Earth's subterranean, the habitat of most interest is in some ways just as remote as a distant planet. Unable to visit the habitat themselves, they have had to be satisfied working in the laboratory with the soil or chunks of rock brought up from the depths by their sophisticated drilling and sampling equipment.

Recently, however, a small team of scientists found a way to live out their fantasy by venturing down into one of the deepest human excavations in the world—South Africa's East Driefontein gold mine. Here a series of vertical and horizontal shafts burrows more than two miles (three kilometers) deep into the Earth's crust. The mine has taken decades to construct and is an engineering marvel by any standard. During a typical production shift, more than five thousand miners are underground, blasting out new tunnels, building support structures, and excavating gold-bearing rock.

In the fall of 1998, Tullis Onstott, a geologist from Princeton University, along with a carefully selected team of microbiologists

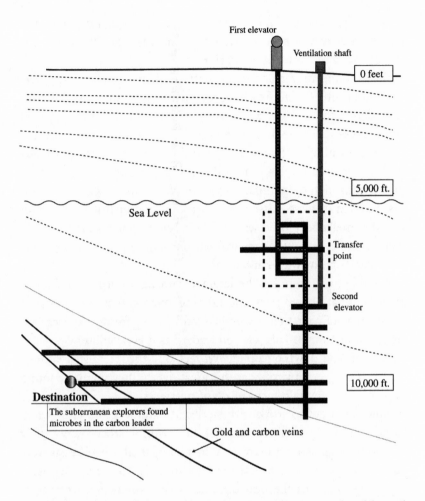

FIGURE 2.1 *South Africa's East Driefontein gold mine, explored by soil biologists who found abundant microbial life within the steamy rocks at a depth of two miles. Illustration by Tamara Clark, adapted from © 1999 Nenad Jakesevic* (Discover *Magazine*).

and other scientists, joined the miners in their dark underground realm for several weeks. On the first day, the researchers decided to head immediately to the deepest and most recent excavations, where the contamination from surface microbes would presumably be minimal. To get there required a two-stage, hair-raising ride in surprisingly wobbly metal cage elevators. On their way down, the researchers could feel the pressure build and the temperature rise as they penetrated deep into Earth's crust—deep enough to experience for themselves the heat generated by magma upwellings and the radioactivity of the planet's interior.

By the time they reached the bottom shaft complex, they were sweating profusely and reaching for their water bottles. At this two-mile depth, the rock surfaces were a steamy 140 degrees Fahrenheit (60 degrees Celsius). Despite the oppressive heat and humidity, the area was bustling with activity. The researchers had to shout to hear each other over the clamor of drills, high-pressure hoses, and air gushing through the ventilation pipes. The beams of light from miners' headlamps were crisscrossing through the dust-filled darkness, and the smell of explosive was in the air. Ignoring the noise, the physical discomfort, and the very real danger of cave-ins, the researchers were nearly giddy with excitement.

A guide took them on the long walk along a horizontal shaft to their destination—a black meandering vein of rock, at points no wider than a finger, that was embedded within a freshly exposed face of the mine tunnel. This "carbon leader," as it is called, is believed to have been laid down by an ancient reef, and the scientists reasoned that if there were microbes at this great depth, they might be found within this potentially rich source of nutrients.

They soon got to work with sterilized hammers and chisels and began removing small sections of the crumbly rock, which they placed in sterilized plastic bags. When all of their sample bags were full, they spent some time snooping around. Although they were planning to return the next day, they were as reluctant to leave as astronauts on their first moon walk. Eventually, when their energy began to fade, they hiked back to the elevator for the ear-popping return ride to the surface.

It was not until months later that they were able to complete the laboratory analysis of their samples. They found that some samples contained much higher populations of microbes than expected— between 100,000 and 1 million per gram. Many of these microbes had unusual metabolisms. For example, some of them "breathed in" iron oxide (rust) as a substitute for oxygen in their respiration process. Others exhaled methane ("natural" gas) as a waste by-product, which was subsequently utilized as an energy source by another important group of microbes sharing the habitat. A particularly intriguing discovery was that some of these creatures were able to utilize metals besides iron, such as cobalt and uranium, in their biological processes. The fact that the highest concentrations of gold are found in the same carbon-rich rock where these microbes reside has led to speculation that the gold has been deposited there as a result of microbial activity. This idea is good news for the scientists, since it justifies more trips to the underground.

NOT ALL EXTREMOPHILES FOUND on our planet live in such exotic locations. Some inhabit common soils and water bodies and rely on their special tolerance for extreme conditions only when necessary. In 1888 the first heat-loving, or thermophilic, bacterium was discovered, floating in the River Seine, along with many other more typical microbes. It was only in the laboratory that its amazing ability to grow at a temperature of 163 degrees Fahrenheit (73 degrees Celsius) was revealed.

The first hint that there might be extremophiles living deep within the Earth came in 1926 when Edson Bastin, a geologist, and Frank Greer, a microbiologist, both of the University of Chicago, found evidence of bacterial activity in groundwater samples extracted from an oil deposit. A more concerted effort along these lines was made during the 1940s and 1950s by Claude Zobell of the Scripps Institute, working in collaboration with Shell Oil Company. He isolated bacteria from oil and sulfur deposits as deep as 12,000 feet (3,700 meters). The pressure at this depth was four hundred times that at the surface, and temperatures were just below the boiling point of water, about 200 degrees Fahrenheit.

During this same period, Russian scientists also reported finding heat-loving bacteria at great depth during oil exploration. Unfortunately, all of this early work was suspect and largely ignored because the researchers had not taken adequate steps to ensure that their samples were not contaminated by microbes at or near the surface during collection. Zobell claimed that he could observe living examples of his bacteria in the laboratory after they had been exposed to temperatures above boiling, but his methods aroused suspicion of possible contamination. Whether or not these suspicions were well founded, most scientists at the time were not ready to accept the possibility that the components of any living cell—its fragile membranes and complex biological molecules—could withstand temperatures beyond about 160 degrees Fahrenheit.

A turning point came in 1965 with the discovery of a bacterium, eventually named *Thermus aquaticus,* that was found thriving at a temperature of 176 degrees Fahrenheit in the steaming hot springs of Yellowstone National Park. The results in this case were unequivocal—the bacteria were harvested alive, and their growth at this temperature was convincingly demonstrated in the laboratory. It was not long before a commercial value for this microbe was found. One of its heat-stable enzymes proved to be useful in a high-temperature procedure called the polymerase chain reaction, which today is used by molecular biology research laboratories throughout the world. The discovery of *T. aquaticus* led to a Nobel Prize and hundreds of millions of dollars a year in revenue for the Swiss drug company that sells the enzyme.

The discovery of *T. aquaticus* forced scientists to raise the bar, so to speak, and establish a new maximum temperature threshold for life. The new magic number became 212 degrees Fahrenheit (100 degrees Celsius), the temperature at which water boils. This would most certainly represent an upper limit, because all life as we know it requires water in the liquid state to serve as the solvent for biochemical reactions. But to the shock of the scientific community, this barrier too was soon broken.

In the late 1970s, microbes growing at 230 degrees Fahrenheit were found thriving in sediments and murky waters near hot, volcanic vents at the bottom of the ocean floor. Life in this environment was possible because the pressure was so high at these depths that water could remain in the liquid state even though temperatures exceeded the sea-level boiling point temperature. So scientists were forced to revise their thinking again. Although the assertion that liquid water is essential for life was upheld, the assumed upper temperature limit was clearly too low because no one had been quite imaginative (some might say crazy) enough to guess that we would find living creatures near the dark, high-pressure, deep-sea hydrothermal vents.

The hunt for extremophiles on *land*—in the deep subterranean— did not begin in earnest until the early 1980s. The primary motivation was to search for microbes that might help in the cleanup, or bioremediation, of some of our buried toxic wastes. By this time, our groundwater pollution problems had become severe in some places, and we were running out of options. We knew that microbes found in some topsoils were capable of converting toxins to harmless products, but it was unclear whether these—or any other microbes, for that matter—were active at the depths where major drinking-water aquifers were located. No one had looked before.

The effort was led by James McNabb, John Wilson, and their colleagues at the U.S. Environmental Protection Agency (EPA), along with collaborators from both academia and industry. Their first challenge was to overcome the sample contamination problems that had plagued earlier researchers. A special drilling apparatus was designed that could bore through dirt and rock while preventing entry of outside air or mud into the core sampler. In addition, a sterile protocol even more stringent than that of a hospital surgical unit was established for handling the samples when they arrived at the surface.

The program was successful almost immediately. A multitude of diverse microbes were discovered, some of which have proven useful for bioremediation purposes. Over the years, this program also has provided useful information on the management of toxic sites; for example, it has shown that aerating or adding fertilizers encourages the activity of de-toxifying microbes.

As it became apparent during the 1980s that microbes with unusual appetites were plentiful at significant depths, the U.S. Department of Energy (DOE) became interested. The DOE had some of its own messes it was concerned about—buried nuclear wastes from the cold war era—and many of these wastes were at greater depths than the groundwater aquifers being investigated by the EPA. DOE researchers hoped their own efforts might help to avert environmental disaster. A geologist and manager at DOE, Frank J. Wobber, also saw research on pollution management as an opportunity to expand fundamental research on subterranean life. He obtained enough DOE funding to establish the long-term Subsurface Science Program, which was designed to go much deeper than anyone had gone before. In addition to specific bioremediation goals, the program had a broad objective: To seek out deep-dwelling life forms and examine their activities. Wobber had no trouble finding scientists from university, government, and private laboratories around the country willing to offer their expertise and facilities.

Again, a first priority was developing the appropriate methodology. The DOE team improved on the drilling and sampling techniques so that they could successfully gather clean samples from depths of several thousand feet below the surface. In one of their initial efforts

in 1987 near a nuclear materials facility in South Carolina, they discovered an abundance of unique microbial life forms flourishing in
the rock at high temperatures, without sun or oxygen, at a depth of
1,500 feet. Since then, numerous other projects at other sites have
found organisms at depths as low as 9,000 feet (nearly three kilometers) and at temperatures often greater than 160 degrees Fahrenheit.
Overall, the DOE program has not only provided useful information
for the management of buried nuclear wastes but significantly
advanced the field of subterranean biology.

IT SEEMS THAT NO MATTER where we poke a hole in our Earth these
days, we find abundant subsurface biota and many unique life forms.
Single-celled organisms, mostly bacteria or the bacteria-like archaea,
predominate, but at some of the relatively shallow depths, protozoa
and higher life forms such as fungi have been reported. Some scientists now speculate that the total biomass of deep subsurface life may
exceed the total biomass of living material on the Earth's surface!

As I stroll among the massive oak, maple, and beech trees in the
Ellis Hollow forest near my home, it is hard to accept the possibility
that there could be a more substantial quantity of living matter within the microscopic pore spaces of the soil and rock beneath my feet.
To wrap our minds around the concept, we must begin with the
recognition that sun-warmed surface life is confined to an extremely
thin layer, like the skin of an apple, compared to the vast, habitable,
core-warmed volume below.

Thomas Gold, a planetary scientist at Cornell University, did the
math on this a few years ago and published the results in a journal
of the National Academy of Sciences. He assumed a depth limit for
the potential subsurface habitat of only about three miles (five kilometers). This assumption was based on a conservative estimate of
the upper temperature limit for life, and on the knowledge that as
one bores deep into the Earth's continental crust, temperatures
increase by about 72 degrees Fahrenheit per mile (25 degrees
Celsius per kilometer). He then assumed that just 1 percent of the
total pore space available within the soil and rock of this layer
would be filled by living organisms. Putting these estimates together, he calculated an underground biomass of 2×10^{14} (200 trillion)
tons. This estimate is indeed higher than our estimates of all surface
flora and fauna. If this biomass were spread out over the entire land

surface of the planet, it would create a layer of living material over four and a half feet (one and a half meters) thick.

Because our subterranean research is still in its infancy, few scientists are willing to go quite as far out on a limb as Gold. One problem in estimating the subsurface biosphere's magnitude is that many of the microbes retrieved from the core samples don't grow when we try to culture them in the laboratory. It is generally assumed either that they die in transit or that we simply don't know how to grow them, but another possibility is that some may have been dead to begin with. Most scientists prefer to wait until we've poked more holes and improved our sampling and handling methods before attempting to quantify life in the underground. Despite these uncertainties, there is an overwhelming consensus that we have discovered a new "deep, hot biosphere," as Gold was the first to call it, that was not recognized even two decades ago.

MANY SOIL MICROBIOLOGISTS today are focusing on the mechanisms, at the molecular level, by which extremophiles are able to live at life's edge. This work is essential not only to expanding biotechnology applications but to advancing our knowledge about evolution and the origin of life. One of the fundamental questions is: How do the heat-loving extremophiles manage to thrive at temperatures that would cook most organisms like a poached egg?

The upper temperature limit for life remains an unknown. Scientists have been proven wrong so many times now that many are reluctant to hazard a guess. When pressed, some would now cite 300 degrees Fahrenheit as the new theoretical maximum. But no one is really sure how many tricks—evolutionary adaptive mechanisms— Mother Nature might have up her sleeve. Of course, it is important to keep in mind that if life originated in a hot subsurface environment, as many now believe, it is we who have adapted to cooler temperatures, not the other way around.

The high-temperature record holders at the moment are all microbes in the domain Archaea. One of these, *Pyrolobus fumarii,* has an *optimum* temperature for growth of 223 degrees Fahrenheit, can continue growing at temperatures of 235 degrees, and has even been reported to withstand autoclaving for one hour at 250 degrees. Its growth slows as soon as temperatures dip below boiling, and at temperatures below about 195 degrees it essentially begins to "freeze" to death. This microbe makes the extremophile bacteria, with typical optimum growth temperatures in the 175 degrees Fahrenheit (80 degrees Celsius) range, seem like wimps. In contrast to the archaea, no bacteria have yet been found that can continue growth when temperatures exceed the boiling point of water.

The archaea as a group are quite distinct at the molecular level from bacteria and other organisms (a subject we examine in greater detail in chapter 3), but some surprisingly simple modifications in the structure of important biological molecules often seem to cause profound effects on temperature sensitivity. The melting point of cell membranes, for example, can be altered simply by changing the degree of saturation (the number of double as opposed to single bonds between carbon atoms) in the lipid molecules that they are made of. This is the same phenomenon observed when margarines with different percentages of saturated fats melt or become hard at different temperatures.

The three-dimensional structure and function of some protein-enzymes can be made resistant to heat by making a single substitution in their long chain of amino acids. One speculation is that this would change the coiling of the molecule just enough to create more cross-links—chemical bonds—at points where the strands come close together. More cross-links would make the protein more stable at high temperatures, where molecules become more active and can jiggle apart. The DNA of some heat-tolerant organisms appears more coiled than in non-heat-tolerant relatives, and these additional twists may in a similar way create more cross-links and more heat stability.

THE ABILITY TO "BREATHE" or respire in an environment without oxygen gas is a must for most subterranean extremophiles because their habitats have very little, if any, oxygen. We have long known about microbes that are "anaerobic"—able to survive without oxygen. Louis Pasteur first observed them in the mid-nineteenth century. It is therefore one aspect of extremophile physiology that we understand reasonably well.

As obligate aerobes ourselves, unable to survive more than a few minutes without our precious oxygen, we tend to marvel at anaerobes for adapting to environments without it. But again, it is worth keeping in mind that it is probably we, not they, who have adapted. The atmosphere of early earth, prior to the evolutionary "invention" of photosynthesis, contained very little oxygen, and most, if not all, of the first inhabitants of our planet were thermophilic anaerobes.

The proliferation of photosynthetic organisms that began about 2.8 billion years ago resulted in a rapid rise in atmospheric oxygen, which is a by-product of photosynthesis. Not only do anaerobes have no need for oxygen gas, but for many of them it is a highly toxic, destructive substance because they do not produce natural antioxidants (as we and other aerobic creatures do). Therefore, as oxygen levels climbed, surface anaerobes had to adapt to the change or find a way to escape. The surviving obligate anaerobes today are found in places like the deep subsurface, oxygen-depleted soil and water, and the digestive tracts of humans and other animals.

From the perspective of the anaerobic microbes, the oxygen by-product of photosynthesis represents one of the first "pollutants" produced by living organisms. In a more objective sense, it is a classic example of how the living affect their own environment and influence the pattern of evolution on a global scale.

Virtually all organisms, oxygen-lovers and oxygen-avoiders alike, are capable of respiring. The energy released during the process of respiration is used to do the work of the living—for example, synthesizing complex biomolecules for growth and maintenance of a living state, or attacking a prey. The function of oxygen—or its substitutes in the case of anaerobes—is in the oxidation, or burning up, of high-energy (high-calorie) fuel molecules, such as sugars. Respiration is not unlike what goes on in a car engine, only there the fuel molecules being oxidized are petroleum hydrocarbons and the energy released does the work of moving pistons.

Biological respiration begins with the stripping off of high-energy electrons from sugar or other fuel molecules. These electrons are eventually passed along through an electron transport chain of special molecules found within most living cells. At each step in this carefully controlled process, some energy is released as the electrons move down an energy cascade, from a state of high to low energy. The energy released along the way is what the organism captures, stores (usually in the form of adenosine triphosphate [ATP]), and later uses for its vital life functions. The presence of oxygen, or

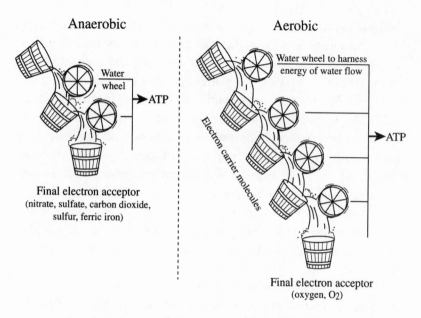

Anaerobic   Aerobic

FIGURE 2.2   *A water cascade that captures the energy released as water flows from the highest to the lowest bucket. This process is analogous to the energy cascade of respiration, by which all organisms capture the energy released from high-energy sugar molecules and store it as adenosine triphosphate (ATP), which is later used to drive the biochemical reactions essential for life. Humans and other aerobic organisms use oxygen as the final electron acceptor, whereas anaerobic organisms use a variety of alternatives. Because oxygen has a very low potential energy, aerobic organisms can usually gain more energy per sugar molecule than anaerobic organisms. Illustration by Tamara Clark.*

another strong electron acceptor molecule, at the very bottom of the cascade is essential to keeping the process moving along.

It might be useful to imagine a cascade of water falling through a series of buckets, starting several feet above the ground (figure 2.2). The distance between the top bucket and the lowest bucket will, of course, determine the total distance the water can fall, and therefore how much work the system can perform. In this model, the water in the initial state, before the top bucket is tipped, represents the high potential energy of the electrons of the original fuel molecules. We know that if we tip the bucket and let the water flow, the energy of the flowing water can be used to do work (for example, to turn a small water wheel to run an electric generator). The lowest bucket at the very bottom of the cascade represents oxygen or some alternative final electron acceptor in respiration. If the final bucket is just a short distance below the top bucket, the water does not flow as far and less energy is obtained than if the final bucket is far below the top bucket.

Of all molecules, the diatomic oxygen gas that we breathe (O_2, two atoms of oxygen joined together) is an ideal final electron acceptor for respiration because it sits very low on the energy cascade. Because the outer electron orbits of the oxygen atoms are only partially filled, they are a strong magnet for other electrons. Compared to other options, oxygen allows a greater "fall" in the energy cascade of the electron transport chain, and a greater release of energy.

But the anaerobes, which evolved before a substantial amount of pure oxygen was available, rely on other options. Ferric iron (an iron atom short one electron), for example, serves as the final electron acceptor for the aptly named iron-reducing bacteria. Other anaerobes use oxygen-containing molecules such as nitrate, sulfate, or carbon dioxide as a substitute for pure oxygen. Thus, life without a supply of oxygen (O_2) gas is quite possible. However, the anaerobic respiration pathways release less net energy than the aerobic approach (figure 2.2). As a result, anaerobic organisms often have a slower overall metabolic rate than aerobic ones and may grow more slowly.

EXTREMOPHILES ARE ALSO remarkable for their ability to find nourishment within their barren, rocky habitats. The most basic needs of all life forms are carbon and energy. We obtain these basics from sugars, fats, and other organic compounds in the plant- and animal-based foods that we eat. Some extremophiles also live off of dead plants in a way: They utilize ancient buried plant life, often in the form of oil, coal, or other hydrocarbons, as a food source.

But it turns out that there are other sources of organic carbon buried deep within the Earth that did not originate from the photosynthetic process of plants. Long ago when our planet was first forming, many of the meteors that bombarded the primitive Earth were a type known as carbonaceous chondrites. These contain organic forms of carbon (molecules with both carbon and hydrogen atoms), as well as nitrogen and sometimes traces of water. Although collisions between the Earth and these types of meteorites are rare today, a recent impact occurred near the town of Murchison, Australia, on

September 28, 1969. The Murchison meteorite was analyzed and found to contain not only organic carbon but several amino acids. Carbonaceous chondrites like the Murchison meteor may have been important in the very origin of subterranean life three and a half billion years ago, in addition to being a source of nourishment for some extremophiles today.

The most amazing of the extremophiles are the lithotrophs, or "rock-eaters," which live off of the rock itself. They obtain their carbon from carbon dioxide gas in a process that is similar to photosynthesis in some ways. However, unlike plants, lithotrophs are in the dark, so they must find an alternative to solar energy to power the uptake of carbon (and to power other life functions). It is only recently that we have discovered that the lithotrophs derive their energy by stripping off electrons from the atoms of inorganic minerals in the surrounding rocks, or from hydrogen atoms of hydrogen gas in the environment. This is a unique and amazing talent, one that allows lithotrophs to survive completely independent of the sun, organic food sources, and surface life.

One of the best studied of the subsurface lithotrophic microbial ecosystems (called SLiMEs, for short) was found in the early 1990s in the Columbia River Basin in the northwestern United States. This bizarre microbial community is embedded in a crystalline basalt rock aquifer thousands of feet below the surface. The anaerobic lithotrophs at the base of the dark food chain obtain their carbon from carbon dioxide gas, and laboratory studies suggest that they are fueled by the hydrogen gas produced when ancient water in the pore spaces reacts with iron-silicate compounds in the rock. They are methanogens that produce methane (natural gas) as a by-product of their metabolism, just as we exhale carbon dioxide.

The discovery of the Columbia River SLiME has global implications because we know that much of the Earth's continental crust is composed of similar basalt rock. If it is verified that hydrogen production from basalt-water reactions can provide enough energy to support an entire community of microbes, this will indicate a widespread "terrestrial energy" source (an alternative to solar energy) for subsurface ecosystems. Even if it turns out that these ecosystems depend on some trickle-down of nutrients from the surface, the fact that this common type of underground rock environment is populated supports the notion that the subsurface biomass of Earth is very large indeed.

THE RECENT DISCOVERIES of Earth's subsurface microbes, with their exotic metabolisms, have caused the U.S. National Aeronautics and Space Administration (NASA) to overhaul completely its plans for future space missions. The lithotrophic ecosystems provide a model for the existence of contemporary life on Mars because basalt rock, liquid water, and bicarbonate (dissolved carbon dioxide gas) are believed to be present within the Martian subsurface. And we know from physical evidence collected by the *Viking Landers* that Martian soils contain abundant sulfur and ferric iron compounds, possible substitutes for oxygen in respiration.

Recently, procedures developed to study life in deep Earth proved very useful in evaluating a rare Martian meteorite found in Antarctica for signs of ancient life. Initially, researchers were almost certain that microscopic wormlike indentations (figure 2.3) were fossils of subsurface Martian bacteria. Soil scientists and microbiologists eventually concluded, however, that these were too small and probably were caused by chemical crystals of some sort, not living organisms. Most tests for signs of life in the rock have not turned up positive, but some feel that small grains of a mineral called magnetite contained in the rock are similar to grains produced by certain subterranean Earth microbes.

Although the evaluation of the Martian meteorite may never be conclusive, many scientists now believe that microbial life on other planets is a real possibility. It's exciting to think about. Clear evidence of extinct microbial life on another planet would be enough to both thrill us and throw us into an existential dither. But in our space explorations to date, we have been looking for life in all the wrong places. Our expensive and sophisticated space probes have literally just been scratching the surface, when the real action may be below, perhaps miles below, the surface.

NASA has recently established the National Astrobiology Institute, both to study life in extreme habitats here on Earth and to plan the details of the search for life in future space missions. The interests of scientists from different disciplines—geology, astronomy, soil microbiology, evolution—are merging around the study of both extreme habitats on Earth and the potential for life beyond our planet. As a

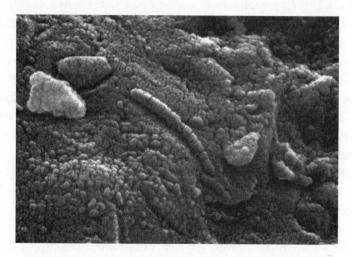

FIGURE 2.3 *The Martian meteorite, discovered in Antarctica, that some have thought may contain microscopic fossils of ancient life (shown in lower, magnified view). The methods developed for identifying signs of life in the deep Earth have proven very useful in the evaluation of this very special extraterrestrial rock. Courtesy of Johnson Space Center, NASA.*

result, the next couple of decades are likely to be filled with many exciting discoveries of life in the underground, and perhaps elsewhere.

TONIGHT SOME *HOMO SAPIENS* somewhere on the Earth (maybe you?) could step outside his or her home, gaze up at the stars, and ask the same question our ancestors must have asked thousands of years ago: Are we alone? We still don't have the definitive answer. But owing to the imagination and perseverance of a handful of researchers exploring the deep Earth, we have some fascinating new information and new clues on how best to approach our search for an answer.

We now recognize that much of the universe is potentially habitable and has all the ingredients for the evolution of life, at least a microbial life similar to the early life of our planet. Stephen Jay Gould, the noted evolutionary biologist, on learning of this fact, put it this way: "We all realize in honest moments that bacteria rule the earth—so why deny them the universe as well?"

Moreover, the subsurface research efforts of the past two decades have turned some of our long-held notions about life on our own planet upside down. It now appears that the habitable space and life forms on Earth may be more vast, by several orders of magnitude, than we ever imagined. We must learn to accept the fact that the sun-warmed surface we are so familiar with is just one of the stages on which life performs its unique feats. Within the context of life in the universe as a whole, the Earth's solar-powered surface may be an almost insignificant component, not center stage.

3

⟡

SHAKING THE TREE OF LIFE

> *Biology, like physics before it, has moved to a*
> *level where the objects of interest and their*
> *interactions often cannot be perceived through*
> *direct observation. And, as in the case of physics,*
> *biology's "subatomic" (subcellular) level is rich*
> *in information, rich in understanding, and rich*
> *in beauty.*
>
> —Carl Woese, Proceedings of the National
> Academy of Sciences (1998)

Late one evening about a quarter-century ago, in a dimly lit laboratory in Urbana, Illinois, a middle-aged scientist sat crouched over a lightbox that illuminated a large sheet of translucent photographic film. Imprinted on the film were rows of dark bands that represented the nucleotide sequence of genetic material that had been isolated from several microbes. The bluish glow from the lightbox filled the room, casting giant shadows on the walls and revealing the face of the scientist. His brow was wrinkled as he focused intently on various details of the film. He lifted his head momentarily and shook it as if in disbelief, rubbed his eyes, and then looked again.

The bar code–like pattern exposed on the photographic film was the culmination of many days of tedious preparatory work. Each row represented RNA fragments from a different organism, and by quantifying the similarity in the location and width of the bands in each row, he could estimate the genetic similarity between the organisms. In fact, he was repeating an analysis he had performed some days earlier. He couldn't believe the results the first time, but now here it was again, the shocking reality staring him in the face. He had checked

53

FIGURE 3.1 *Dr. Carl Woese in his laboratory during the late 1970s, at about the time of his discovery of a new domain of life. Courtesy of the University of Illinois.*

and double-checked his work during all aspects of the procedure. This was not some experimental artifact caused by a mix-up in the chemicals used, an accidental switching of samples, or some other error. The results, if they could be confirmed by additional tests, could mean only one thing—this would be one of the most important discoveries of the twentieth century. He had identified not just a new microbial species but an entire new kingdom, or superkingdom, of organisms!

The scientist was Dr. Carl Woese (pronounced "woes") of the University of Illinois, and the year was 1976. In reality, the discovery unfolded over many days, nights, and weeks. The microbe that shook his world, and eventually initiated controversy and revolution in biology, was considered at the time to be nothing more than an obscure type of bacteria known as a methanogen. As mentioned in chapter 2, methanogens get their name from the fact that they produce methane, or natural gas, as a by-product of their metabolism. It was not known when Woese first studied them, but it is now believed that most of the natural gas deposits buried within the upper mile or two of the Earth's crust have been produced by methanogens. They are also the soil organisms that produce the combustible marsh gas that sometimes hovers over wetlands, rice paddies, and other vegetated areas with waterlogged and oxygen-depleted soils.

What Carl Woese conclusively established in 1976 was that, although the methanogens look like common bacteria under a microscope, genetically they are as distinct from bacteria as bacteria are from plants or animals. In fact, on a genetic basis, the methanogens have less in common with the other bacteria than a redwood tree or fungus has with you or me. If plants, animals, and bacteria are to be considered separate kingdoms, then so must the methanogens.

As Woese expanded his analyses, he soon found that not only the methanogens but many other supposed bacteria were also in the unique genetic category he had discovered. He began referring to the new category as a "domain" and gave it the name Archaebacteria, for "ancient bacteria." Later the name would be changed to simply Archaea to indicate more clearly the uniqueness of methanogens from bacteria and other forms of life. Woese recognized that these findings would shake our concept of the evolutionary "tree of life" down to its roots. What he could not foresee at the time were the personal and professional battles he would have to fight to gain acceptance and understanding of his revolutionary discovery.

I FIRST MET WOESE in the fall of 1998. I had arrived in Urbana on a Sunday afternoon, although our meeting wasn't scheduled until the following morning. I decided to try calling him right then to confirm the time and get directions to his campus office. I only had a work number, but I had a hunch he would be there. Sure enough, he picked up the phone. Despite nearing retirement age, he was taking advantage of a quiet Sunday to catch up on work. As I already knew by his steady stream of publications, he was by no means slowing down.

The next morning I got up early and found my way to campus. On the lower level of the building that housed the Microbiology Department, I stopped for a moment to look at a large hallway display dedicated to Woese as a recipient of the prestigious

Leeuwenhoek Medal, named after Antony van Leeuwenhoek, a pioneer microbiologist of the seventeenth century. I then continued on upstairs to Woese's office, which was actually a small converted laboratory. Much of the bench space held old, antiquated laboratory equipment—perhaps items he did not have the heart to throw away. There were stacks of papers, journals, and books everywhere and a few strategically placed computer monitors, keyboards, and printers. As I entered the room, I could see an older gentleman with thick white hair, short-cropped, leaning back in a swivel chair, his feet up on the lab counter, crossed at the ankles. He looked very much at home; it had to be Woese. The chair with its worn dark green upholstery looked just like one in my office; all land-grant colleges must have the same source. Woese was contemplating something on the computer.

Here is someone on the short list for the Nobel Prize, I reminded myself. I thought about how very different it is to visit a scientist at the top of his or her profession and, say, a politician or business leader at the top of their game. There was no penthouse view, no leather chairs, no large desk made of exotic woods, and no wet bar (unless the old lab sinks with leaky faucets could serve that purpose). Whereas a leading politician or Fortune 500 CEO might have been sporting an Armani suit, Woese was wearing old tennis shoes, loose-fitting khaki pants, and a simple flannel shirt with rolled-up sleeves. I liked it, the understatement, because it indicated a lack of pretense that is common in the sciences.

As is often the case with revolutionaries, Carl Woese entered the field whose paradigms he would challenge—biology—with background in another discipline. His undergraduate training during the 1950s was in physics at Amherst College in Massachusetts. He crossed the bridge to biology some years later by earning a doctorate in biophysics at Yale University. After graduate school, a postdoctoral research project revealed to him for the first time the molecular wonders of the microbial world and the secrets that world might hold for unraveling the origin of the genetic code. After brief periods of employment with General Electric and the Louis Pasteur Institute in France, he landed a tenure-track professor position in the Microbiology Department at the University of Illinois in 1964. Finally, with the breathing room provided by the academic freedom of a university environment, Woese could roll up his sleeves and get down to serious work on the questions that intrigued him.

From the beginning, Woese's major interest has been the origin and evolution of life's most important molecules—the DNA (deoxyribose nucleic acid) and RNA (ribose nucleic acid) that make up the genetic code. The double-helix DNA provides the master copy of an organism's genes, and RNA, a single-stranded version of DNA, translates the genetic code into life's essential processes, beginning with the synthesis of the protein-enzymes that catalyze life's biochemistry. He recognized that an essential first step would be to build a more complete and accurate tree of life, one that encompasses the early evolution of the incredibly diverse microbial world. By identifying the present-day microbes that are the most direct descendants of our most ancient ancestors, he was bound to gain insight into the mother of all cells and the origin of the genetic code itself. It was clear to Woese that the existing tree, emphasizing plants and animals, was artificially skewed toward large, recently evolved surface organisms like ourselves, and so was of little use to him. Shaking the tree of life was just a means to an end, although, as we shall see, it led him to serendipitous discovery, controversy, and career jeopardy.

AN IMPORTANT TURNING POINT for Woese came in 1965 when he read a paper entitled "Molecules as Documents of Evolutionary History" in the *Journal of Theoretical Biology*. It was written by one of the pioneers in quantum chemistry and molecular biology, Linus Pauling, and a colleague, Emile Zuckerkandl. They had been gathering data on the amino acid sequence of biologically important protein molecules for many years. They had begun to notice that when they compared the same protein isolated from different species, the similarity of aligned sequences of amino acids of the proteins was correlated with the amount of evolutionary time that separated the species. Organisms that evolved at about the same time showed nearly identical sequences, while those that evolved at very different times had noticeable differences. These proteins were like a "molecular clock" because they accumulated random changes in their amino acid sequences over evolutionary time. These changes were apparently neutral in that they did not affect the function of the proteins and so got carried along, harmlessly, from generation to generation. When

Woese read that, it confirmed his plan for determining evolutionary histories of the bacteria. Only he would eventually use the nucleotide sequence of genetic material—RNA molecules—rather than the amino acid sequence of proteins, as his molecular clock.

This breakthrough was a recognition that expensive fossil-hunting expeditions, the kind that make such great *National Geographic* covers, are not the only, or even the best, approach to exploring the biological history of life on Earth. Investment in more powerful electron microscopes was not the answer either. Woese and just a handful of others at the time were convinced that within every living cell, at a level beyond the view of microscopes, are clues to our evolutionary past, tucked away in the structure of long, chainlike molecules such as proteins and genes. This approach could not even have been imagined earlier because scientists did not have the techniques for examining the structure of proteins or genes in detail. Woese's plan was to use the newly emerging tools of molecular biology to reach back in time beyond the oldest fossils, to the period when all life was microbial and our most ancient ancestors roamed the planet. Woese would not need to travel to exotic lands to seek out our past; he would do all of his digging in a modest laboratory in Urbana, Illinois.

Woese decided that a small subunit of a type of RNA called ribosomal RNA (rRNA) would be the best molecular clock for his purposes. Ribosomal RNA gets its name from the fact that it is associated with cellular structures called ribosomes that are part of the protein-building machinery of every cell. The particular subunit Woese selected is involved in the synthesis of protein-enzymes that no organism can do without. Thus, it is found in all creatures, from bacteria to begonias, from mushrooms to humans. Using rRNA would allow Woese and others to compare all of Earth's genetic diversity on the same terms and construct a truly universal tree of life. Similar to the changes in the amino acid sequence of the proteins studied by Pauling and Zuckerkandl, the rRNA undergoes random neutral changes in nucleotide sequence that serve as a reliable counting mechanism—the "tic-toc" of evolutionary time.

In the early days, Woese worked in almost total anonymity, ignored by most of the scientific community. Many of those who did pay attention considered him a crackpot who was using an excruciatingly tedious technique that could never answer the big questions he claimed to be interested in. His first step was to isolate the rRNA subunit from cells. Today, with rapid automated equipment, an entire 1,500–1,800 nucleotide rRNA subunit can be sequenced in a couple of days, but when Woese began in the late 1960s, doing so took half a year or more. Therefore, instead of trying to sequence the entire subunit, he used enzymes to break it up into fragments that were about twenty nucleotides long, then sequenced just some of those. This shortcut relied on the fact that statistically it would be highly unlikely that the nucleotide sequence of any fragment more than about six nucleotides long would repeat itself within the same rRNA subunit. Although it would be ideal to know the entire nucleotide sequence, this was not necessary to make comparisons between organisms.

Criminologists rely on this same statistical approach when examining DNA evidence from a crime scene. It would be completely unrealistic to sequence entire DNA molecules, which can be tens of thousands of nucleotides long. (Conducting such an analysis would be similar to conducting another human genome project.) Criminologists rely on the fact that when the sequence of just several fragments of DNA from the crime scene matches exactly the sequence from the same regions of a suspect's DNA, it is "highly probable" that the DNA found at the scene was left by the suspect.

Woese was able to compare analogous fragments of rRNA from any two organisms and to quantify their relative evolutionary age and degree of relatedness based on the proportion of nucleotides that matched up in sequence. From this he was able to construct simple dendograms, or "trees," and determine which organisms belonged on the same branch or twig and where the important branching points were located. For the first time, the "invisible" microbes, which so dominate the underground world and embody much of the genetic diversity and living biomass on our planet, were on an equal footing with multicellular creatures on the tree of life.

Woese labored away in near isolation from the 1960s through the 1980s, while his laboratory shelves became jammed with boxes of the large film sheets containing genetic information for hundreds of organisms. Woese was one of the few scientists in the world who was capable of interpreting these films as "bar codes" representing nucleotide sequences and evolutionary relationships. Gradually, a new universal tree of life began to emerge. It was filled with surprises, some of them small, and others revolutionary.

During my visit with Woese, he took me into a large room lined with shelves from ceiling to floor that had been completely dedicated to the storage of these film sheets, thousands of them, now of historical significance. I was awed by this monument to the hours, weeks, and years of relentless pursuit of a scientific objective—the search for pattern in the relationship between organisms. This sight brought home to me the fact that leading a scientific revolution takes much more than genius. The dedication, stamina, and tenacity of a bloodhound are equally important attributes.

PRIOR TO THE "WOESIAN REVOLUTION," our tree of life was essentially an eye-of-the-beholder version of reality: it was based primarily on what creatures look like, and what we could guess their ancestors looked like from the fossil record. Our evolutionary tree had advanced surprisingly little from the time of the ancient Greeks.

Aristotle, in the third century B.C., described a *Scala Naturae*, or "Ladder of Life," which was a hierarchy that began with inanimate matter at its base, followed by plants, animals, and finally, of course, man at the top (figure 3.2). Almost two thousand years later, in 1735, Carolus Linnaeus published his masterpiece of taxonomy, *the Systema Naturae* ("Natural System"), which has as its two great branches the same plant and animal kingdoms Aristotle described. Linnaeus's important contribution was his hierarchical classification scheme, which is still used today; it divides each of the kingdoms into class, order, genus, and species.

Antony van Leeuwenhoek's discovery in the seventeenth century of single-celled microbial life forms complicated things. Were they plants or animals? Most biologists and taxonomists took the easy way out and simply ignored Leeuwenhoek's work and the microbes until, in the nineteenth century, Louis Pasteur demonstrated the important role they play in causing disease. After that, microbes could no longer be ignored. The problem was (and still is) that most living microbes and their fossils appear as nondescript rods or spheres, preventing accurate classification. Even with the aid of powerful electron microscopes, the incredible diversity of the microbial world is not well revealed by visual images alone.

Rather arbitrarily, it was decided to put the larger, motile, single-celled organisms, named protozoa, into the animal kingdom, and the relatively immobile fungi and tiny single-celled bacteria into the

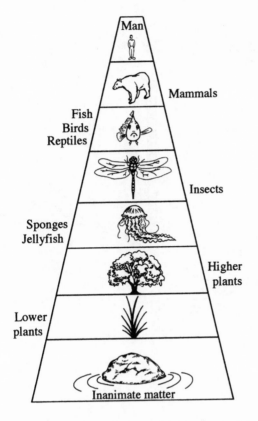

FIGURE 3.2 *The "ladder of life" proposed by Aristotle in the third century B.C. Illustration by Tamara Clark.*

plant kingdom. This classification is in fact what I was taught in high school in the 1960s, even though by then many scientists had decided to lump the protozoa and bacteria together into a third kingdom of their own. When I entered the University of California as a biology major a few years later, I learned the very latest dogma of the scientific community—a five-kingdom classification system proposed by Robert Whittaker of Cornell University in 1969 (figure 3.3). It raised the protozoa, bacteria, and fungi each to the status of individual kingdoms, alongside the old Linnaean favorites of animals and plants.

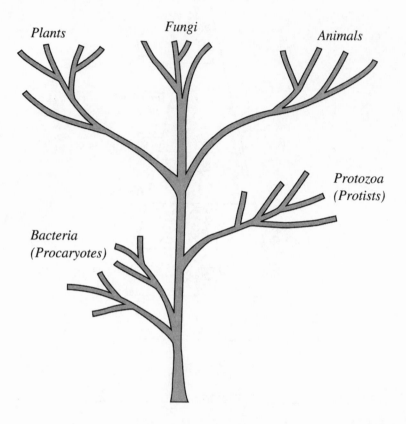

FIGURE 3.3 *The five-kingdom tree of life originally proposed by Robert Whittaker. Illustration by Tamara Clark.*

By that time, the detailed comparisons of organisms made with powerful scanning electron microscopes had revealed that all of Earth's life forms could be grouped into two "superkingdoms" based on cellular structure: The eukaryotes, which have cells with a well-formed nucleus, and prokaryotes, whose cells lack a nucleus. Within the five-kingdom scheme, all multicellular plants, animals (including humans), fungi, and the single-celled protozoa are within the super-kingdom of eukaryotes; only the bacteria are prokaryotes.

This is where things stood when Woese arrived on the scene. Woese was not satisfied with the five-kingdom tree. He knew that the prokaryotes, the bacterial branch, represented most of the evolution-ary history of life on the planet, and that their living members had the metabolic diversity to survive in a wider range of ecological niches than the other four branches. Bacteria and their relatives have been evolving for at least three and a half billion years, while the multicellular creatures emphasized in the five-kingdom tree have been around for less than one billion years.

A tree based primarily on visual characteristics would never do justice to the genetic diversity of the prokaryotes or the unicellular organisms that were at the base of the other branches. Some advances in prokaryotic classification had been made in the twentieth century by grouping bacteria according to the types of nutrients needed to culture them in the laboratory. This approach was seriously constrained, however, by the fact that even with special nutrient enrichment techniques, only about 1 percent of the estimated total diversity of microbial life forms could be cultivated.

So Woese pursued his molecular approach. One by one, he isolated the rRNA of individual bacterial strains and compared fragments for distinctions in nucleotide arrangement. During his first ten years of research at the University of Illinois, Woese gathered enough rRNA data on about sixty types of bacteria to begin publishing their genealogies—that is, the shape of the prokaryote branch. Occasionally he would dabble with the members of the other four branches of the five-kingdom tree, the eukaryotes. What became apparent when he began making comparisons was that within the bacterial branch there are sub-branches that differ as much from each other as plants differ from animals. In other words, if the difference in rRNA nucleotide sequence between plants and animals was to be used as the yardstick to define separate kingdoms, he had evidence that within the bacterial branch alone there are several separate "kingdoms" of prokaryotes.

This was mind-boggling enough, but Woese was in for an even bigger surprise. It all started one day in 1976 when a colleague down the hall, Ralph Wolfe (no relation to the author), supplied him with some colonies of methanogens. Not much was known about the methanogens at the time, except that they appeared to be bacteria, they often inhabited subsurface soils, waters, and other places deficient in oxygen, and they produced methane gas as a by-product of their metabolism. Wolfe was one of the few well-established microbiologists who believed in Woese's approach, and he was curious as to where the methanogens might fit in the bacterial genealogy that Woese was constructing.

Woese put the methanogen sample through his rRNA sequencing mill. As described earlier, when he examined the film that resulted, the sequences for methanogens did not match up with anything he or anyone else had ever seen for a bacteria. And the sequences were also distinct from those of all the eukaryotes—the protozoa, fungi, plants, and animals. For Woese, one of the few who could interpret and fully appreciate the rRNA sequence data, it was as startling as stepping into the backyard and seeing a new bizarre creature that was clearly neither plant nor animal.

Any scientist would be thrilled at discovering a new species to add to our understanding of Earth's biodiversity, but Woese had unexpectedly dredged up an entire continent of new life forms, a new superkingdom. For the next several months, Woese put in even more hours at the lab to confirm his results. He examined other methanogens, and based on the rRNA data, they also turned out to be in the unique group he eventually named the Archaea.

Day by day the evidence accumulated, and soon it was abundantly clear to Woese that all life on Earth could be divided into three primary superkingdoms, or "domains" as they are now referred to: Bacteria, Archaea, and Eukarya. (The Eukarya domain encompasses the former kingdoms of plants, animals, fungi, and protozoa.) These domains have unique "signature" nucleotide sequences in certain parts of their rRNA subunit to establish that they represent the deepest, most fundamental, branches in the universal tree of life.

Within a year of the initial discovery, Woese and Wolfe published their results in the *Proceedings of the National Academy of Sciences*. Their discovery of the archaea did not go unnoticed by the popular press, and in November 1977, it was front-page news not only in Woese's hometown paper, the *Urbana News Gazette*, but even in the *New York Times*.

YOU DON'T HAVE TO BE a rocket scientist to comprehend the scope of the Woesian revolution. You don't even have to be a biologist. And in contrast to the quantum revolution in physics, you don't need mathematical training in Lagrangian or wave functions to fully appreciate this one. The Woesian revolution can be conveyed with a single, compelling image—the universal tree (figure 3.4).

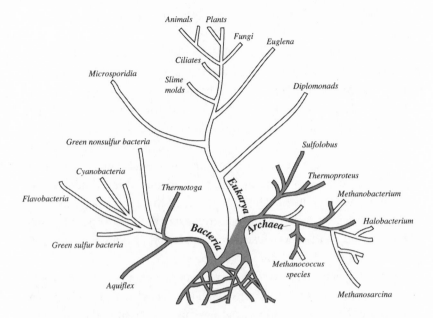

FIGURE 3.4   *The universal tree of life developed by Carl Woese and others using rRNA nucleotide sequence data. The shaded area represents heat-loving microbes. Illustration by Tamara Clark.*

What shocks the socks off most people almost immediately is that the visible diversity of life we see all around us, the multicellular plants and animals, is represented by only two small twigs on one branch, the eukaryotic branch, of the new universal tree. It clarifies how our overreliance on visual evidence has for thousands of years warped our perspective on the evolution of life on our planet. Most high school and introductory college textbooks on biology today continue to perpetuate this thinking by emphasizing the plant and animal kingdoms. The rRNA analyses tell us that within each of the three domains of life are dozens of other kingdoms. And most of those kingdoms, most of Earth's genetic diversity, is microbial.

The prokaryotes, previously thought to be a single branch of primitive creatures within a five-kingdom tree dominated by large multicellular life forms, are now recognized as representing fully two-thirds of Earth's genetic diversity—the Archaea and Bacteria domains. There is greater diversity and evolutionary distance, by several orders of magnitude, within the new domain of Archaea discovered by Carl Woese than exists among the plants, animals, and fungi combined.

Throughout the 1990s the pace stepped up as more rRNA of new organisms was sequenced and the tree was filled in. By 1998 more than five thousand organisms had been classified in this way. In 1996 the complete genome (not just rRNA fragments) of one methanogen, *Methanococcus jannaschii,* was sequenced and the results were reported in *Science.* Parts of the *M. jannaschii* genome were similar to bacteria, but other parts were more similar to eukaryotes. Overall, the results verified that archaea occupy a unique third domain, even though they *look* like bacteria. Since then, the complete genomes of several other archaea have been sequenced, and all of these findings tend to support the conclusion reached much earlier by Woese in his analysis of the rRNA fragment data alone.

The universal tree provides a molecular genetic approach to the study of the origin of life on Earth. The fact that heat-loving thermophiles have the oldest evolutionary history—that is, they are at the base of the universal tree (see figure 3.4)—is weighty evidence in support of the hypothesis that life originated in a high-temperature habitat such as the deep subsurface or within sediments near oceanic volcanic vents. The rRNA data suggest that all three domains—the Archaea, Bacteria, and Eukarya—arose from a common community of primitive life forms long ago, rather than one branch arising from another. This is a radical departure from a belief held for centuries, including much of the twentieth century, that the multicellular eukaryotes represented "higher" life forms that evolved from the more primitive prokaryotes.

It now appears that the three domains branched apart long ago and have for the most part evolved independently. However, near the base of each domain within the universal tree, the relationships get messy because these most ancient single-celled creatures are capable of "laterally" transferring genetic material to distantly related organisms, even across domains. At this primitive level, genes are swapped with the abandon of a 1960s "free love" festival, although it is all done in a G-rated asexual fashion. In this "you are what you eat" realm, loose genetic material released by the damaged cells of one species can be engulfed like food by active cells of another species and incorporated into their genome.

As we gradually fill in the base of the tree over the next decade or two, it may come to resemble a network more than a simple branching pattern. Even with the powerful tools of molecular genetics, the precise location of the root of the tree may remain a mystery.

Many of the archaea are thermophilic and among those direct descendants of our earliest ancestors that now inhabit extreme environments. However, archaea are also being discovered in other environments, some of them cold rather than hot, and others not very extreme at all. For example, scientists have found a very diverse and numerous group of archaea thriving in the cold ocean waters in Antarctica. In the North Atlantic, archaea have been found among the bacterial communities devouring the sunken luxury liner, the *RMS Titanic*. These microbial communities are extracting iron from the steel superstructure and producing huge iron-containing "rusticles," some of them meters in length, that hang from the sunken ship. Yet another example resides in topsoils. Until very recently, topsoils were not generally considered a likely place to find archaea, but it now appears that this conclusion was based primarily on our inability to culture them. Utilizing new molecular methods, scientists examining soil samples from regions as diverse as Finland, the Amazon, and Wisconsin have found evidence of archaea living among the earthworms, insects, and microbes in upper soil horizons. The precise ecological role of the archaea found in soils and cold ocean waters is at this time largely unknown.

CARL WOESE BROUGHT THE STUDY of evolution into the molecular age and, in so doing, brought the microbes of the underground into the Darwinian fold. In 1977, when Woese first went public with his findings about the methanogens, he knew he had made a contribution that most scientists can only fantasize about. He had, after all, discovered a third domain of life! But what happened next—or rather, what did *not* happen—was discouraging.

After the initial few weeks of attention and newspaper reports, the requests for interviews quickly dwindled. As the months passed, the struggle to find funding to continue his work did not improve, and there was no flood of eager graduate students clamoring for a post in Woese's laboratory. The worst part of that period, according to Woese, was that most microbiologists simply ignored the mountain of evidence for a three-domain tree of life that he had so painstakingly put together. They refused to believe that this scientist, working on his own for years examining tiny bits of bacterial rRNA, was really on to something. Some openly criticized Woese's work, scoffed at his conclusions, and warned his supporters that they were jeopardizing their own careers by associating with him.

During my visit, I asked Woese whether he felt that there is something wrong with the scientific process, something we should fix, considering that the battle to convince the scientific community of his revolutionary ideas is still being fought in some quarters. To my surprise, he answered, "It's appropriate that science move cautiously on matters as profound as this. Corroboration from other laboratories just took time. Now that we have faster automated methods, and we're sequencing the entire genome of organisms, things should move more quickly. Maybe some of the puzzles and inconsistencies can be resolved." Still, he conceded, the field might have advanced more quickly had his colleagues been more open-minded.

In retrospect, Woese recognized that part of the problem was his isolation. He loved his work, but he did not get much satisfaction from attending scientific conferences. With his background in physics and his molecular perspective, he spoke a different language than others involved in microbiology and evolutionary studies at the time. He would rather be in the lab, sequencing the rRNA for a new organism, than socializing with fellow scientists and lobbying their support for his interpretation of the data.

Another, somewhat related problem was that, at least in the beginning, only a small number of scientists were doing similar work and could comprehend the rationale of his approach or the implications of his results. Data from other labs to confirm or refute what he was finding were hard to come by. To Woese, the interpretation was obvious, and the data overwhelming. It should have spoken for itself. But it did not.

Fortunately, Woese's credentials and scientific methods were impeccable, and a slow but steady stream of his publications made it through the peer review process. He gained a handful of well-respected and influential supporters, including Norman Pace, an evolutionary biologist at Berkeley, Otto Kandler, an influential German microbiologist, and of course Ralph Wolfe, his University of

Illinois collaborator. This small support group stood by him, often putting their own reputations on the line.

The cold shoulder from the scientific community did little to dissuade Woese. Stubborn and self-confident by nature, he dug in his heels. He read Thomas Kuhn's *The Structure of Scientific Revolutions* and gained some comfort from the fact that his struggle was not unique in the history of scientific advance.

When Woese mentioned this, I thought of the parallels between his story and that of Antony van Leeuwenhoek. In the seventeenth century, while Galileo was searching the sky for planets and stars, Leeuwenhoek, a cloth merchant by trade, was exploring droplets of pond water for microscopic "animalcules" and "wretched beasties," as he called them. Leeuwenhoek had a handful of supporters, most notably the famous British naturalist Robert Hooke, but for the most part he worked in anonymity and his findings received a lukewarm, at times even hostile, response. His difficulties in being accepted may in part have been due to the fact that he was isolated from much of the scientific community. He was not a bona-fide member of the academic club. Another problem was that Leeuwenhoek's lenses (which he ground himself) and technique were so superior that no one could duplicate his results. Leeuwenhoek took it on the chin gracefully. In a letter to a friend he wrote:

> Among the ignorant, they're still saying about me that I am a conjuror, and that I show people what does not exist; but they're to be forgiven, they know no better. . . . Novelties oft-times aren't accepted, because men are apt to hold fast by what their Teachers have impressed upon them.

Luckily for us, Leeuwenhoek pursued his work and documented his findings. After his death, bacteria—those "wretched beasties"—would not be seen by human eyes again for at least another century. Finally, in the nineteenth century, others came along who were able to match his skills with a microscope and confirm his observations, and we began to recognize the potential significance of a microbial world.

Like other scientists who before him have had the fortune, or misfortune, of being at the helm of a scientific revolution, Woese has had to take the heat. But no scientific revolution can be credited to a single man or woman. This one, a revolution still in progress, is no exception. Carl Woese owes a serious debt to the likes of Linus Pauling and other pioneer molecular biologists who preceded him. And by the late 1970s, Woese was no longer working alone. Others were independently recognizing the advantages of the molecular approach to the study of microbial evolution. Otto Kandler, who was making his own discoveries about the uniqueness of the methanogens by analyzing their cell walls, was as convinced as Woese that the archaea represent a third unique domain of life thriving on our planet.

Gradually, during the 1980s, the tables slowly turned, and the number of microbiologists who belittled Woese's efforts began to dwindle. The rRNA of several hundred organisms, representing all three of the major domains, was characterized. By the end of the decade, most scientists had at least come to accept that archaea are a unique life form, although many continued to dispute that the archaea deserve their own branch on the evolutionary tree. Woese, once shunned by many microbiologists, had become one of their leaders, even hailed as a hero by some. Woese's universal tree of life is now considered dogma among microbiologists, and the number of skeptics in other fields is dwindling as well. Virtually all of the scientific community now acknowledges the genetic uniqueness of the archaea, and most would agree that rRNA analysis has become an important tool for clarifying evolutionary relationships.

DURING THE PLANE RIDE HOME after my visit with Woese, I reflected on the modern scientific process. We are seeking truth, a deeper understanding of the world around us, but no one wants to get involved in a wild goose chase. Most of the criticism that Woese has faced, and continues to deal with, is based on the legitimate concerns of dedicated scientists, not on petty jealousies. Peer review of grant proposals and publications and many other, more subtle barriers have been established to prevent one renegade scientist from leading us all over the cliff and into the dreaded Abyss of False Theories. This is a good thing, of course, but for the scientist with a new perspective on an old problem, however, the process of convincing colleagues that he or she is right can not only be grueling and painfully slow but pose a serious career risk.

Woese recalled for me some of the factors that had kept him motivated all those years. Primarily it was the work itself, he said, and confidence that he was making progress in tracing the genetic code back to its roots. But there were some unexpected and pleasant surprises too. For example, in 1980 his German colleague Otto Kandler invited him to the first international conference on the archaea in Munich, and when Woese arrived, he was treated like royalty. He found that, thanks largely to Kandler's considerable influence, his ideas were being enthusiastically accepted in much of Western Europe. And when the moment came for Woese to give the talk he had prepared, a full choir and brass orchestra broke into celebratory music as the guest of honor approached the podium. Kandler had arranged this reception as an antidote to the emotional toll he knew Woese was suffering owing to criticism and lack of recognition in the United States.

Just a decade after this event, in 1990, worldwide recognition, at least within the field of microbiology, was no longer an issue. In that year, Carl Woese was flown to Amsterdam to receive microbiology's highest honor, the Leeuwenhoek Medal, awarded by the Royal Netherland Academy of Arts and Sciences. This award was initiated in 1875, the bicentenary of Leeuwenhoek's discovery of "animalcules." The award is not given out lightly or often. There have been only a dozen recipients in the past 125 years. Louis Pasteur received the medal in 1895. One is inclined to imagine that none on the prestigious list of recipients would have pleased Leeuwenhoek more than Carl Woese.

I asked Woese whether receiving "the Leeuwenhoek" was his most gratifying moment, the one that most made him feel it was all worth it. He thought for a moment and then shook his head. "Here, let me show you something." He walked to a nearby office shelf and pulled down an old 1991 edition of *The Biology of Microorganisms*. This is a widely respected textbook in microbiology, and one that Woese himself has often referred to over the years in its many editions. He opened the book, and there, within the front cover, was a complete diagram of his three-domain universal tree of life, based on rRNA data. "That," he pointed at the page, "that did it."

Part Two

LIFE SUPPORT FOR
PLANET EARTH

4

OUT OF THIN AIR

*In denitrification and nitrogen fixation, we see
the visceral biological forces at work and the
intimate union of life with the fluxes and pools of
the global metabolism.*

—Tyler Volk, *Gaia's Body:
Toward a Physiology of Earth* (1998)

René Descartes, the famous French mathematician
and philosopher of the seventeenth century, became a legend when
he uttered: "I think, therefore I am." A modern evolutionary biologist might come to a similar conclusion, but with this bit of logic:
"We struggle, therefore we exist." The history of all life forms on
Earth today is one of a struggle for existence, as Darwin put it, in the
midst of scarce resources and a changing environment. You probably have felt the pinch yourself at times. Without this struggle, there
would have been no evolution beyond primitive self-replicating biomolecules, and we would not be here to wonder at the grand mystery of it all.

No struggle has been more acute than the search for nitrogen, a
scarce but essential element. Every living cell on Earth requires it,
from independent microbes like the archaea, bacteria, and protozoa,
to the interdependent cells that are members of complex, coordinat-
ed colonies, like our bodies. Along with carbon, oxygen, and hydrogen, nitrogen is very much the "stuff of life." Most of the nitrogen
within living organisms is in the form of amino and nucleic acids, the
basic building blocks of proteins and our genes. As discussed earlier,
the formation of these macromolecules was essential to the very origin of life on our planet. Protein-based enzymes catalyze fundamental biochemical reactions, and the nucleic acid sequences of our DNA

and RNA direct the synthesis of proteins and provide a means for self-replication and evolution.

It is specialized creatures of the underground that control much of the global nitrogen cycle and keep it churning fast enough to support us all. These species evolved slowly, over the course of hundreds of millions of years, as the biosphere expanded and nitrogen demand increased. It is fortunate that life on Earth, through genetic mutations, is constantly experimenting with new metabolic processes for tapping into potential sources of energy and nutrients. The more scarce a particular element is, the more likely it is that organisms with new variations on ways of acquiring it will succeed and pass on their ability to subsequent generations. If it were not for this tendency, we would have consumed ourselves out of existence long ago. Over evolutionary time, what was inaccessible or considered "waste" becomes "food" for new life forms. From this perspective, you could say that scarcity has served as the fertilizer for the evolutionary tree of life.

The struggle for existence is commonly portrayed as a combative, head-to-head competition between species. However, as we shall see in this chapter and others to follow, life in the underground is often based on cooperation between organisms. Although the modus operandi of genes and individuals is "selfish," nature has evolved a number of complex symbiotic relationships between species in which cooperation leads to a mutually beneficial outcome. One of the most important symbioses on Earth is that between certain species of plants and a type of soil bacteria with the unique ability to capture nitrogen out of thin air.

IT IS A CRUEL IRONY of Mother Nature that nitrogen is such a scarce resource, because the fact is that we are literally bathed in the stuff. About 78 percent of every cubic yard (or meter) of the air that surrounds us is diatomic nitrogen gas (N_2). There are seven tons of this gaseous form of nitrogen over every square yard of Earth surface. Take a deep breath. Most of what is filling your lungs is nitrogen. The problem is that, unlike oxygen (O_2), which we are able to absorb from the air and into our bodies by its reaction with the hemoglobin of our blood, N_2 is chemically inert, or nonreactive. The two nitrogen atoms in the N_2 molecule are held tightly together by an unusually strong triple bond, and only a small number of microbial species have the chemistry to break this bond apart. Millions of N_2 molecules are gushing into and out of our lungs with every breath, but not one of them is assimilated. If the atmospheric N_2 gas that surrounds us were the only source of this precious element, our species and most others on the planet would soon be extinct—sort of like dying of thirst in the middle of an ocean.

For all but a handful of soil and marine microbial species, then, the immense supply of nitrogen in the air may as well be on Jupiter or Mars. No plant, animal, or fungus can process N_2 gas as a nitrogen source. And to make matters worse, very little nitrogen exists elsewhere, other than even less accessible forms of nitrogen buried deep in sedimentary and igneous rock. Over 99 percent of the Earth's nitrogen, excluding that buried deep within the rocky crust, exists in the atmospheric pool as N_2 gas. All of the nitrogen found in the soils, seas, and living things makes up the remaining 1 percent.

We and all other life forms on the planet, including most microbes, depend on a special group of prokaryote (bacteria and archaea) species called "nitrogen-fixers" to convert N_2 gas into something the rest of us can use. The evolutionary "invention" of nitrogen fixation ranks with photosynthesis (carbon fixation) as one of the cornerstone events in the history of life on Earth. While photosynthesis created a mechanism for the biosphere to tap into solar energy and the carbon in carbon dioxide gas for the first time, nitrogen fixation allowed access to the huge storehouse of nitrogen contained in the atmosphere. Many believe that photosynthesis evolved first, perhaps three billion years ago, followed by nitrogen fixation a billion years later. Taken together, these two biological processes greatly increase the carrying capacity of the Earth to support life by unlocking resources that otherwise would be unavailable.

The first, and most difficult, hurdle that must be overcome by the microbes that fix N_2 gas is to break the unusually strong chemical bond that holds the two nitrogen atoms together. Once freed, these nitrogen atoms can be combined with hydrogen, carbon, or other elements to build amino acids, and eventually proteins, or other biologically important molecules. When the nitrogen-fixers die and decompose or are consumed, the nitrogen enters the food chain in organic molecular forms that can be utilized by the rest of us.

Essentially all of the nitrogen contained within the proteins and genes of our bodies has been funneled through the nitrogen-fixing microbes at one time or another.

Although most soils contain some nitrogen-fixers, the ability to capture N_2 gas from the atmosphere is a relatively rare attribute. Among the many tens of thousands of bacterial species in the soils and seas of the world, there are probably no more than two hundred species of free-living nitrogen-fixers. Somewhat more common are those that live within the roots of their host plants, providing the plants with nitrogen and receiving in return the products of photosynthesis (carbon- and energy-rich sugars).

All nitrogen-fixers of the world, whether free-living or symbiotic, rely on the same enzyme—nitrogenase—to orchestrate the conversion of N_2 gas into ammonium (NH_4, a much more usable molecule made up of one nitrogen atom combined with four hydrogens). Nitrogenase is a giant among enzymes, both in the literal sense—it is huge and complex—and in its significance for global biogeochemistry. For those of us who tend to fret about nature's fragile balance, it can be a bit disconcerting to learn that there are probably no more than several pounds (a few kilograms) of this precious enzyme on the entire planet. The entire world's supply of nitrogenase could fit into a single large beaker or bucket! Lose this and life on Earth as we know it would come to a screeching halt.

We have recently identified more than twenty bacterial genes involved with the synthesis and control of nitrogenase, and its detailed structure has been determined through X-ray crystallography and other techniques. The long twisted chains of atoms it is composed of are arranged like a bowl of spaghetti, or a ball of yarn after a cat has been playing with it for a while. But its physical shape, each twist and turn of the chains of atoms, is not a random affair. Those twists and turns, along with various electrical charge configurations, help to hold the N_2 molecules in place while other molecules, such as water, are carefully brought into proper position to stimulate chemical interactions. Nitrogenase is composed of two giant proteins that physically separate and come back together eight times, over the course of 1.2 seconds, to convert one molecule of N_2 to one molecule of ammonium. Most chemical reactions occur in nanoseconds. A duration of 1.2 seconds, thousands of times longer than most biochemical processes, is almost unheard of and reflects the difficulty of nitrogen fixation. Attempting to fix N_2 without nitrogenase as a catalyst is a challenge for even the most sophisticated of modern laboratories.

The nitrogenase enzyme greatly lowers the energy requirement for the breaking apart of the incredibly strong triple bond holding the atoms of nitrogen together in N_2. Nevertheless, compared to other biochemical processes, nitrogen fixation is very "expensive" with regard to the energy it eats up. Because it is such an energy-demanding business, free-living nitrogen-fixers are at a disadvantage. They must scrounge their environment for carbohydrates or other high-energy molecules to drive the process, whereas symbiotic nitrogen-fixers get these inputs from their plant hosts.

IT WAS NOT ALL THAT long ago—one hundred years or so—when we first learned of the symbiotic association between soil microbes and plants that is so essential to meeting the nitrogen needs of life on land. This discovery, like so many others, was to some extent serendipitous. Two obscure German agricultural scientists, Hermann Hellreigel and Hermann Wilfarth, uncovered the phenomenon while conducting a series of routine greenhouse experiments to determine the nitrogen requirements of various crops. Their curiosity got the better of them when they noticed that crop species in the legume family, such as beans, peas, alfalfa, lupines, and vetch, often did much better than other crop species when grown in nitrogen-deficient soils. The responses of legumes to nitrogen fertilizer additions were also much more variable than the responses they observed for other crops. By measuring the actual nitrogen content of the tissues of the plants, they determined that legumes require just as much nitrogen as the other crops but mysteriously seem able to meet their requirement regardless of the amount in the soil.

The value of legumes as rotation crops had been recognized for thousands of years, but no one before Hellreigel and Wilfarth had been able to link their value specifically to nitrogen. Publius Vergillus Maro, aka Virgil, the famous Roman poet who lived from 70 to 19 B.C., recommended rotating with legumes in *The Georgics,* his ode to workers of the land:

> Sow in the golden grain where previously
> You raised a crop of beans that gaily shook
> Within their pods, or a tiny brood of vetch,
> Or the slender stems and rustling undergrowth
> Of bitter lupine...
> Thus will the land find rest in its change of crop,
> And earth left unplowed show you gratitude.

The same advice for farmers, albeit less poetically stated, appears in modern literature distributed by presumably "cutting-edge" land-grant institutions, such as Cornell University where I work. Of course, now we know quite a bit more about the how and why of legume effects, and with this information, we can fine-tune recommendations for farmers.

It took Hellreigel and Wilfarth many months of tedious experiments to conclude that certain soils contained *something* that was not nitrogen but somehow had a positive, nitrogen-like fertilizing effect on certain legumes. Was it some chemical substitute for nitrogen that they could not detect? A more definitive experiment found that when peas (a legume) were grown in steam-sterilized soil, they lost their advantage and developed yellow leaves and other symptoms of nitrogen deficiency, just like the nonlegumes. This experiment, by proving that it was something *living* in the soil that provided the legumes with the nitrogen, suggested some type of symbiosis.

In 1886 Hellreigel presented a paper summarizing three years' worth of experiments to the Fifty-ninth Conference of German Scientists and Physicians held in Berlin. The publication of this work created an immediate sensation because concern was growing at the time that farmers would soon run out of an adequate supply of nitrogen fertilizer to feed the rapidly expanding human population. Many scientists working at more sophisticated facilities than Hellreigel and Wilfarth's agriculture experiment station were a bit suspicious. If symbiotic nitrogen fixation was as widespread as Hellreigel and Wilfarth suggested, why had no one noticed it before? They rushed back to their laboratories to try to repeat the results.

To the surprise of many, the results were easy to confirm. In retrospect, it is clear why the more sophisticated laboratories had not made the discovery—they seldom used actual soil for growing the plants, preferring instead a soil-less growing medium with which

FIGURE 4.1 *Root nodules containing nitrogen-fixing bacteria on the roots of a soybean plant. Photo by Joseph Burton.*

they could better control the chemical status of the root environment. As it turned out, this is one time when the reductionist approach and the use of the simplest "model system" led researchers astray.

In 1888 a young and ambitious German scientist named Martinus Biejerinck, who was already well on his way to becoming an important pioneer in the field of soil microbiology, isolated for the first time bacteria in the genus *Rhizobium* growing on the roots of legume plants. These nitrogen-fixing bacteria were found within the small nodules, often about the size of a corn kernel, that characteristically form in clusters along the roots of legumes and other nitrogen-fixing plants (figure 4.1). With this discovery, all of the pieces of the puzzle began to fall into place.

It was soon proven that the *Rhizobium* bacteria produce ammonium from N_2 gas that they gather from the air spaces in the soil around them, and that this ammonium is used as a nitrogen source by the legume plant hosts. This explained why legumes require less nitrogen fertilizer than other crops. It also became clear why legumes are so beneficial as rotation crops: Some of the symbiotically fixed nitrogen is released into the soil in usable organic forms when the plants die and decompose, fertilizing the soil for the plants and other organisms to follow.

Legumes continue to be of tremendous importance in agriculture as high-protein crops for use as food or animal forage and in rotation with other crops as "green manures." Today the legume seeds sold to farmers are routinely inoculated with strains of *Rhizobium* that have been selected or genetically engineered for maximum nitrogen fixation ability. Leguminous trees play an important role in maintaining soil nitrogen levels in some natural ecosystems, such as the mesquite in the Mojave Desert, the acacias in many semidesert and savanna regions, and several hardwoods unique to tropical rain forests. The most widespread and ecologically important nitrogen fixers on land are bacteria in the genus *Rhizobium,* but in some forests of Scandinavia, the northwestern United States, and other temperate regions, nitrogen-fixing bacteria in the genus *Frankia* dominate, providing nitrogen through their association with alder, bayberry, and other tree species.

In the past decade, we have learned a great deal about symbiotic nitrogen fixation at the molecular level and the steps by which bacteria and plants identify each other as "friend" rather than "foe." *Rhizobium* bacteria contain more than twenty-five nodulation genes, which are usually found on special small rings of DNA called plasmids. For their part, legumes and other host plants also possess genes specific to the nodulation process. Think of what the host plant is faced with. It must screen the tens of thousands of bacteria surrounding its roots and choose those few that are "the good guys"— *Rhizobium* bacteria—that it will allow to penetrate its tissues. This screening process must override the defense mechanisms against parasitic infection that plants have acquired over many millions of years of evolution.

The first step in establishing a friendly relationship between the *Rhizobium* bacteria and the legume is taken by the plant. When the plant is at a particular developmental stage and the environmental cues are just right, certain plant genes are turned on, leading to the production of special compounds called flavonoids. These flavonoids exude from the roots, attracting *Rhizobium* bacteria and triggering several nodulation genes of the bacteria to turn on. The flavonoids produced by different legume species are unique: They attract only certain species of nitrogen-fixing bacteria and trigger only their genetic program.

Once the process has been initiated, the *Rhizobium* bacteria begin to produce complex sugars that cause the tiny root hairs of the plant to curl and allow the bacteria to penetrate through a kind of tunnel that winds its way among the root cells. When the bacteria reach special host cells of the root, they invade the cells, the cells enlarge, and the bacteria-plus-cell complex, called a bacteroid, is coated with a special semipermeable membrane produced by the plant. This membrane directs traffic, allowing some of the ammonium produced by the bacteroid to leak out while allowing carbohydrates and essential nutrients needed by the bacteroid to leak in. The final step is the construction of a large nodule of plant tissue around the bacteroids.

The key enzyme in the nitrogen fixation process, nitrogenase, is destroyed if exposed to oxygen. Each type of nitrogen-fixer has therefore devised its own mechanisms for protecting nitrogenase from oxygen contamination. In the *Rhizobium*-legume symbiosis, large oxygen-absorbing macromolecules of leghemoglobin are produced and bathe the bacteroid. Scientists were amazed when they discovered that production of the "legheme" portion of leghemoglobin is controlled by *Rhizobium* genes, while the "globin" portion is coded for by plant genes. Two species from separate domains sharing the know-how and labor to synthesize a complex macromolecule—it's the ultimate in symbiosis! Leghemoglobin is unique to *Rhizobium*-infected legumes, but as the name implies, it is very similar to the hemoglobin of our blood. Like hemoglobin, leghemoglobin contains iron and turns a deep red color when oxygenated. Active nodules on the roots of legumes often have a blood-red color.

All of this intricate nitrogen-fixing infrastructure and activity is expensive. It has been estimated that *Rhizobium* bacteria and other symbiotic nitrogen-fixers often burn up about 20 percent of the carbohydrates generated by the photosynthesis of their plant hosts. Therefore, while plants benefit from the symbiosis by gaining nitrogen, they pay a heavy price. It is not surprising that nature has worked out mechanisms that shut down nitrogen fixation when it is not needed. For example, some of the genes controlling nodulation

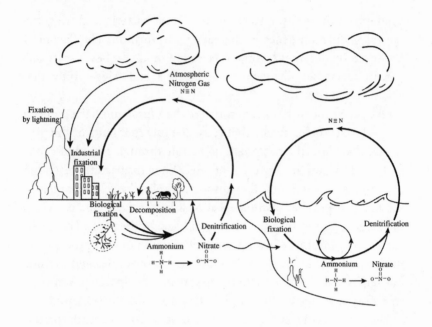

FIGURE 4.2 *A simplified diagram of the global nitrogen cycle. Illustration by Tamara Clark.*

and nitrogenase synthesis are turned off when high amounts of ammonium are already present in the soil.

THE BIOSPHERE HAS EXPANDED greatly during the past three and a half billion years. As already mentioned, this expansion would never have been possible without the concomitant dramatic increase in biological nitrogen fixation, which increased the overall carrying capacity of the Earth. However, the nitrogen demands of life on Earth today are so great that nitrogen fixation supplies only 10 to 20 percent of the annual requirements. The rest is supplied by *reusing* the nitrogen that has previously been fixed and is already a part of the biosphere, soil, and sea nitrogen pools. In this recycling, just as with the initial fixation of nitrogen, soil organisms play a major role (figure 4.2).

Very early on, as the food chain began to evolve, a huge cast of soil creatures, collectively known as decomposers, developed the capacity to obtain their nitrogen by consuming dead microbes, plants, or animals. Today these creatures break proteins and other nitrogen-containing macromolecules down into a simple ammonium form, use some of it for themselves, and leave the rest behind in the soil to be taken up by plants or other microbes. Humans, like other animals, eventually became a part of this internal loop of the nitrogen cycle, obtaining nitrogen by consuming plants or other animals.

If this internal recycling of nitrogen were totally efficient, we would not depend so heavily on the input of new nitrogen from the atmosphere through nitrogen fixation. But it's not. Because much of the nitrogen that enters the biosphere and soil pools leaks out again, a continual nitrogen fixation subsidy is absolutely essential.

The leaks in the internal loop of the terrestrial nitrogen cycle begin with the fact that much of the organic and ammonium nitrogen in soils gets converted to nitrate (a nitrogen combined with three oxygen atoms, NO_3) by the activity of a small but ubiquitous group of microbes. Nitrate, unlike ammonium, is not held very tightly by soil clay particles and so is easily washed away by rainfall into groundwaters and rivers—eventually ending up in the oceans (as depicted in figure 4.2)—before it can be utilized by plants or microbes in the soil.

Much of the soil nitrate that does not get washed away by rains ends up escaping from the land in another way—it is converted back to N_2 gas by a highly specialized group of soil microbes known as denitrifiers. The denitrifiers are most active in flooded soils but are also found in drier environments within the oxygen-depleted interiors of dense soil aggregates. Denitrifiers produce the N_2 gas as a waste by-product of their anaerobic respiration process, which utilizes the oxygen atoms of nitrate as a substitute for the diatomic oxygen gas (O_2) that we breathe.

Globally, the denitrifiers shunt vast quantities of nitrogen from the soils (and seas) back into the atmosphere. If we look at this process as a reversal of all the hard work of the nitrogen-fixers, it might seem like a bad thing for Earth life. But in fact, if there were no denitrifiers to capture nitrate, much of Earth's nitrogen would eventually drain into the oceans in its nitrate form and become unavailable to life on land. Creatures in the ocean could use some of this nitrate, but ocean life is constrained by shortages of iron and other elements and could never use it all.

The denitrifiers are like a cleanup crew that collects much of the nitrate leaking from the internal recycling system on land before it drains into the ocean and is lost for the long term. By converting nitrate back into N_2 gas, the atmosphere is replenished and the external loop of the nitrogen cycle is closed. Without denitrifiers, nitrogen would have a one-way ticket, flowing from the atmosphere to the biosphere and eventually accumulating in the oceans as nitrate. Over the long term, life on our planet is just as dependent on the activity of the denitrifiers as on the nitrogen-fixers.

IN THE EARLY TWENTIETH century, as scientists were just beginning to piece together the nitrogen cycle, a panic came over them. Many warned that the supply of organic and other usable sources of nitrogen, such as manure and mined nitrate deposits, would not be adequately to keep up with the fertilizer requirements for feeding a rapidly increasing human population. These fears were given considerable credence when, in an address to the Royal Society of London in the early 1900s, the prominent British scientist Sir William Crookes painted a bleak picture and warned of mass starvation. The fate of our species was precariously dependent on the activity of those few species of subterranean microbes, the nitrogen-fixers, that could make usable nitrogen out of N_2 gas.

What really got the attention of the general public and politicians, however, was an impending war in Europe. Nitrate is an essential ingredient for trinitrotoluene (TNT) and most other explosives used in warfare. (With the exception of nuclear arsenals, this is still largely true today.) A supply of nitrogen was therefore considered an urgent matter of national security. At the time, the only significant source of nitrate for both agriculture and explosives was found in deposits on some small islands off the coast of southern Chile. In the unusually dry and cold climate of the islands, the guano of nesting seabirds had accumulated as nitrate over tens of thousands of years. The Chilean deposits were being heavily mined, and if no alternative was found, they would be completely depleted in just a few years.

So the race was on during the first decade of the twentieth century to figure out how to do what the subterranean nitrogen-fixers had been doing for billions of years—make ammonium out of N_2 gas. (Relatively straightforward industrial methods for creating nitrate and fertilizer, once ammonium had been acquired, were already known.) The threat of running out of ammonium and nitrate was, in effect, the selection pressure that drove the human species to evolve

FIGURE 4.3 *Fritz Haber, "the man who fixed nitrogen." Haber was awarded the Nobel Prize in Chemistry for this work in 1919. Courtesy of Tom and Maria Eisner.*

the capacity to fix N_2. The best minds were attracted to the problem, and funding from industry and government agencies was made available to them. At the time, Germany had some of the most sophisticated physics and chemistry laboratories in the world, and it was from one of these laboratories that a scientist with the solution stepped forward.

His name was Fritz Haber. To suggest that he was the first chemist to obtain ammonium from N_2 gas is a bit misleading. More precisely, Haber was the first to do it without blowing up his laboratory in the process, and with a yield of ammonium that made his procedure practical for commercial production. A contemporary of Haber's in Germany, the famous physical chemist Walther Nernst, was also tackling the problem, and there was a bit of a competition between the two. Nernst was already famous for coming up with the third law of thermodynamics, among other things. Haber was four years younger and equally ambitious, but less well known at the time. Nernst was a brilliant theoretician, but Haber was more gifted in the practical art of chemistry, and his procedure for ammonium synthesis produced higher yields. (Nernst insisted that Haber's results were theoretically impossible until, some years later, it was found that one of the constants Nernst was using in his calculations was erroneous.)

Haber's ammonium yields continued to increase as he refined his methods, using different metals as catalysts, as well as new types of containers to withstand higher temperatures and pressures. In 1909 he patented his process and made arrangements to begin the first commercial production of ammonium. The nitrogen crisis was soon over, and Haber began to reap the financial and professional benefits of his invention.

The talents of "the man who fixed nitrogen" were directed toward a very different goal during World War I, when Haber was asked to spearhead the German effort to develop chemical weapons. He was a sincere German patriot and agreed, rationalizing his decision with the thought that the fear generated by the threat of chemical weapons would shorten the war and result in less suffering overall. Other scientists were appalled. When in 1919 he was awarded the Nobel Prize in Chemistry for his nitrogen fixation work, some Frenchmen who were offered awards the same year declined them in protest, calling him "morally unfit for the honor."

Despite being misunderstood and shunned outside of Germany, Haber maintained an active laboratory and mentored a following of outstanding young scientists. As the Nazis took power in Germany in the 1930s, however, Haber, like all citizens of Jewish ancestry, found his life turning sour. Max Planck, a high-ranking German physicist, is said to have tried to defend Haber and other Jewish scientists in a conversation with Hitler. Hitler replied, "If the dismissal of Jewish scientists means the annihilation of contemporary German science, then we shall do without science for a few years."

Haber was forced to flee his beloved homeland in 1933. Because of his World War I activities, however, he was still not welcome in many parts of Europe. Within one year of exile, on January 29, 1934, Fritz Haber died in his sleep of a heart attack in Basel, Switzerland. Eighteen years later, in 1952, a memorial plaque was dedicated at the Kaiser Wilhelm Institute in Dahlem, Germany. It states, in part:

> Haber will live in history as the brilliant discoverer of an exceedingly important means of advancing agriculture and the welfare of mankind, who obtained bread from air, and won a triumph in the service of his country and all of mankind.

Today, after nearly a century of trying alternatives, the Haber process remains the only economically viable way to produce synthetic nitrogen fertilizers. We humans still have not come up with a way to fix nitrogen at room temperature as the microbes do with their amazing nitrogenase enzyme. The Haber process requires large amounts of energy, temperatures of about 930 degrees Fahrenheit (500 degrees Celsius), and pressures of several hundred atmospheres (several hundred times air pressure at sea level). Constructing a nitrogen fertilizer plant costs hundreds of millions of dollars—a substantial investment.

Despite the expense and energy consumption of the Haber process, the demand has been so great that the amount of nitrogen fixed industrially in this way has been doubling about every six years since Haber first introduced his method (figure 4.4). In the United States, more than 90 percent of the fertilizer nitrogen needs in most cropping systems are met by man-made nitrogen fertilizer. Today, on a global basis, more nitrogen is fixed by humans with the Haber process than by all natural microbial nitrogen-fixers on land! In other words, we are interfering with the nitrogen cycle in a serious way. But we have done so for a legitimate reason. It has been estimated that, at a minimum, one-third of the human population at any one time is being fed by the Haber process.

Although the Haber process has helped to feed a hungry world, it has simultaneously created one of our most serious environmental threats—excessive amounts of nitrogen pollutants in our soils, air,

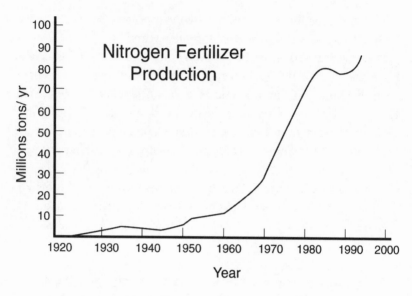

FIGURE 4.4 *Annual rates of nitrogen fertilizer production. Adapted from data at www.fertilizer.org.*

and waterways. This environmental problem could be minimized if we used fertilizer more efficiently and if we recycled the nitrogen-rich organic wastes left behind by the increased populations of humans and livestock animals that are supported by increased nitrogen inputs. Unfortunately, however, only about one-third to one-half of the nitrogen we pour onto our farms, lawns, home gardens, and golf courses each year is taken up by plants. Much of the rest leaks into groundwaters, estuaries, and the oceans as nitrate.

High levels of nitrate in drinking water are toxic to humans—particularly infants—and other animals. Nitrate added to rivers, freshwater lakes, and estuaries throws the food chain out of balance and puts into motion an undesirable process called eutrophication. Nitrate stimulates massive blooms of algae and various aquatic microbes, which clog the waterways, reduce the clarity of the water, and deplete most of the oxygen from the water so that few other species can survive. Eutrophication due to excess nitrogen can turn a pristine bay or marine water inlet into a fish-less, green, mucky mess, dominated by just a few plant and animal species.

The huge increase in nitrogen inputs supplied by humans has accelerated the entire global nitrogen cycle. With a faster cycle and a greater quantity of nitrogen being moved about in each spin of the cycle, more nitrogen has been leaking. It's not just nitrate that is a problem. Recently, scientists have become concerned about a leak involving a gaseous form of nitrogen known as nitrous oxide (N_2O). Nitrous oxide can potentially damage the stratospheric ozone layer, cause "acid rain," and it is a very potent "greenhouse gas"—three hundred times more potent than carbon dioxide. This gas escapes in trace quantities during intermediate steps of the denitrification process and the conversion of ammonium to nitrate. As the populations of the microbes carrying out these conversions have increased (in response to greater nitrogen inputs to soils by humans), the amount of nitrous oxide released to the atmosphere has increased. The historical trend in the rise in atmospheric nitrous oxide concentrations (from about 290 to 310 parts per billion in the past fifty years) has paralleled the rate of increase in human nitrogen fixation activity.

Should we have seen these nitrogen pollution problems coming? In hindsight it is easy to think so, but the fact is that it was not until the latter part of the twentieth century that the nitrate and nitrous oxide problems became obvious. Only a handful of scientists were aware of the potential hazards earlier on, and obviously their voices were not heard. But even if we had been clearly forewarned, would that have stopped us from using the Haber process to keep pace with the food and nutritional needs of a rapidly expanding human population?

Pondering the answer to such questions is probably less important than coming up with solutions. Many are trying to do just that. Agricultural scientists and farmers are working together to come up with ways to minimize nitrate losses into groundwaters, streams, and lakes. The efforts are multifaceted. Initially, the emphasis was on education—raising awareness among farmers that excessive nitrogen fertilizer application is not only uneconomical but potentially harmful to the environment. The education effort has been followed up by a concerted research effort focused on improving the precision of nitrogen fertilizer applications and the efficiency of crop uptake. Many farmers are learning to "spoon-feed" fertilizer to their crops slowly as the crops develop, rather than take the more wasteful approach of applying it all, in luxurious quantities, just before planting. One exciting new technology, already being tested in some areas, is the use of global positioning satellite (GPS) data to create high-resolution maps of the nutrient status of farm soils. This information is then fed into a computer on a tractor that varies the amount of fertilizer applied, foot by foot, as a farmer passes through his fields. Old technologies are being revisited too. For example, significant strides are being made in the genetic improvement of legumes and their nitrogen-fixing bacterial partners, and the use of legumes as "green manure" rotation crops is gaining favor once again. Some farmers are including in their rotations "trap crops" with deep, prolific rooting systems that take up nitrate before it leaches into groundwater. Finally, geneticists are hoping to reduce the nitrogen requirements of important world food crops.

We have reason to hope that the same human ingenuity that has found ways to keep our human population well fed will find solutions to the environmental problems created by our tampering with the nitrogen cycle. We know that our nitrogen-fixing activities of the past century have significantly increased the carrying capacity of the Earth to support life. If we want to sustain that life, and the quality of our natural environment, we will need to come up with better, more efficient ways to manage the nitrogen we fix.

5

〰〰

NEXUS OF THE
UNDERGROUND

*In the days when the "information superhighway"
is the buzz phrase, we would do well to look at our
inventive fungal predecessors who, for four
hundred million years, have already been leading
the communication network of life on land.*

—LYNN MARGULIS, FOREWORD TO *HYPERSEA* BY MARK
AND DIANNA MCMENAMIN (1994)

OVER THE CENTURIES, NATURALISTS AND POETS have often attempted to describe their intuitive sense of a connection between living
things. Science is now revealing how very real and intimate some of
these connections are. If it were possible to gaze down and witness
everything below our forests, grasslands, and other natural ecosystems, one of the most striking things we would see is the vast network
of gossamer-like fungal threads linking the roots of plants of different species. We have only recently learned that the expansion of life
up and out of both the sea and the deep Earth was founded on this
very important symbiosis between plant and soil fungus. It has been
just as fundamental to the evolution of life on our planet as the relationship between nitrogen-fixing bacteria and legume plants. The
roots of almost all plant species in the world today are joined with
these specialized fungi that help them obtain water and nutrients
from the soil (and sometimes from neighboring plants) in exchange
for the carbon- and energy-rich sugars produced by the plant during
photosynthesis. It's a connection that began long ago, when life on
the land surface was just getting started.

THE COLONIZATION OF EARTH'S land surface did not begin in earnest until the early Devonian period, a little over four hundred million years ago. This land invasion was more than three billion years in the making; during that time, the subterranean creatures that would be required to support surface life became established. The first photosynthetic organisms to attempt life on land came ashore from marine environments as rootless, green, algae-like creatures. Needless to say, these sea-dwellers were in for quite a shock when they tried their luck on land. Most quickly shriveled up and died. Over the course of millions of years, however, a few managed to survive by being the first to establish a successful partnership with soil fungi. The fungi functioned as surrogate roots, supplying their algal partners with water and nutrients mined from the underground, while the algae collected solar energy at the surface and supplied the fungi with the products of photosynthesis.

The descendants of those first photosynthetic species gradually evolved into primitive plants with roots of their own. The symbiosis with fungi, however, did not end there. Today, more than four hundred million years after the first tentative union between algae and fungus, you are likely to find descendants of those beneficial fungi growing on almost any plant you yank up by the roots, regardless of where on Earth you might be. In many cases, you would need to use a microscope and special staining procedures to spot the delicate fragments of threadlike hyphae dangling from the roots, but they would be there. These fungi-root associations are referred to as mycorrhizae—from the Greek *mykos* (fungi) and *rhiza* (root). More than 90 percent of the approximately 248,000 species of higher plants on our planet today have found it advantageous to continue this mutually beneficial partnership. In fact, the majority of plant species could not survive in nature without it.

Many of the mycorrhizal fungi are just as dependent on their plant partners as the plants are on them. We are unable to culture some of the most common types in the laboratory, even with our most sophisticated concoctions of sugars and nutrients. Scientists have speculated that these fungi depend on plants not only for a share of the products of photosynthesis but also for some as yet unidentified essential growth hormones.

The fossil record confirms that the first mycorrhizal fungi evolved at just about the same time as land plants. Their unique hyphal

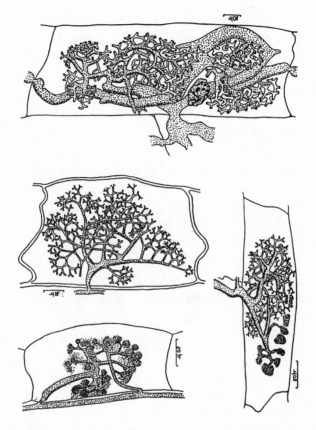

FIGURE 5.1   *Microscopic images of arbuscular mycorrhizae, as drawn by the pioneer researcher I. Gallaud and first published in* Revue Generale de Botanique *17(1905): Plate 4.*

branching structures can be viewed through a microscope. Resembling miniature tree shapes, these structures, called arbuscles—from the Latin *arbor* (tree)—form just within the root tissues (figure 5.1) and are the interface for the exchange of nutrients with the plant host. In 1994 fossils from an important archaeological site called the Rhynie Chert, near Aberdeen, Scotland, were reexamined and found to contain evidence of arbuscles. The rocky layers in which these fossils were discovered are just over four hundred million years old. Along with the fungi, fossil evidence of primitive pioneer land plants was also found.

Genetic analyses have provided additional evidence of the long history of this symbiosis. In the early 1990s, the first sequencing and evolutionary classification of ribosomal DNA collected from arbuscular mycorrhizal fungi confirmed that they originated between 350 million and 460 million years ago, coincident with the estimated time of origin of the first land plants. Today arbuscular mycorrhizae are found almost everywhere except in some arctic regions, and the list of host plant species is very long. Most temperate and tropical non-woody plants, such as grasses, wildflowers, and our most important crop plants, are host to this type of mycorrhizae, as well as woody perennials such as azalea, apple, grape, cedar, maple, ash, and many tropical trees. We have not yet found a single arbuscular mycorrhizal fungus that grows independently of a plant host.

Both the fossil record and other DNA studies reveal that a second major group of mycorrhizal fungi, called ectomycorrhizae (figure 5.2), evolved about 160 million years ago. The ectomycorrhizae are unique in that they do not form arbuscles inside the root tissues, and their silken hyphal filaments extend out farther from the roots (often several yards) compared to arbuscular mycorrhizae. The ecto-mycorrhizal fungi release a plant hormone that causes the growth of short, stubby, branching rootlets here and there along the main roots of the plants they inhabit. These characteristic multibranched rootlets are about one-eighth of an inch long (three millimeters), so they can usually be identified in the field without the aid of a microscope. The ectomycorrhizal roots are composed of a central core of plant tissue completely encased by a dense mat of fungal hyphae. The ectomy-corrhizae are particularly important in many temperate and arctic forests. They form symbioses that are crucial to many economically important timber trees, such as pine and other conifers, oak, beech, chestnut, and birch. The list of their plant hosts is not as long as that of the more ancient arbuscular mycorrhizae, but they have a wide geographic range and are found growing with woody perennials—and even a few nonwoody species—from the tropics to the arctic.

There is a great deal of overlap in geographic range and plant hosts between the ectomycorrhizae and the more ancient arbuscular mycorrhizae. Both types of root fungi can often be found in the same forest, meadow, or crop field, and they may even be growing on the

FIGURE 5.2   *Ectomycorrhizae on pine (above), with the characteristic shortened, thick roots covered by fungal hyphae (courtesy of Ken Mudge, Cornell University); and a magnified view (below) of the thick hyphal mat surrounding a modified root hair (courtesy of Hugues Massicotte, University of Northern British Columbia).*

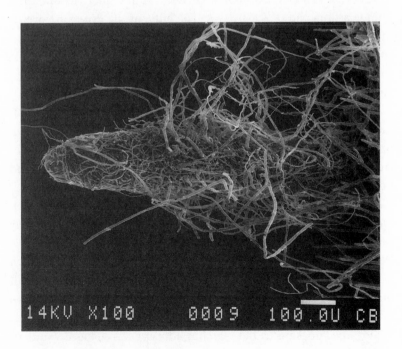

same plant. Maples and poplar trees, for example, are sometimes simultaneously a host to representatives of both major groups of mycorrhizal fungi. In general, the fungal root symbionts are not nearly so host-specific as the nitrogen-fixing bacteria discussed earlier. The mycorrhizal fungi are also much more ubiquitous and abundant in nature than the nitrogen-fixers. Those doing research with mycorrhizae have sometimes found it necessary to not only sterilize the soil to create a "no-mycorrhizae" control treatment but to filter the air entering the greenhouse to prevent free-floating volunteer spores of mycorrhizae from contaminating their experiments.

THE DISCOVERY OF MYCORRHIZAE was, as is often the case, serendipitous. The road to discovery began in the early 1880s when the king of Prussia commissioned one of the world's leading forest biologists, Professor A. B. Frank of the Landwirtschaftlichen Hochschule in Berlin, to study, of all things, truffles. These highly prized delicacies are the below-ground fruiting (spore-forming) body of a rare type of aromatic fungus (*Tuber melanosporum* and *T. magnatum* species) found in some hardwood forests of Western Europe. Then as now, truffles sold at incredibly high prices because of their rarity and their unique aroma and delicate flavor. The hope of the Prussian government was that Frank would develop a way to produce truffles on a commercial scale, like the common (and much less expensive) grocery store mushroom, which is the aboveground fruiting body of another type of fungus (*Agaricus brunnescens*).

Professor Frank failed rather miserably at coming up with a commercial method for the cultivation of truffles. (Incidentally, so have all others who followed him, which is why, if I wanted to buy a pound of truffles today I would have to be willing to pay close to one thousand dollars per pound!) There are two edible species—black (*T. melanosporum*), native to Germany, France, and Spain, and white (*T. magnatum*), found in northern Italy. Both types are still extremely rare. They have been sought after for their culinary qualities since Roman times. The black Perigord truffle reportedly was hunted down in fifteenth-century France by trained muzzled pigs, who could sniff out the smelly fungi. The highly competitive truffle hunters of today, or *truffiers*, as they are called in France, roam their secret forest haunts with trained dogs instead of pigs, but little else has changed. Humans and their trained animals aren't the only truffle hunters—voles and other rodents seek them out by smell and spread the spores as they take the truffles back to their underground burrows.

What Frank did discover launched a century's worth of research, the results of which we are only now beginning to appreciate fully. "We should all fail so nobly," as Michael Allen, one of today's leading mycorrhizal researchers, put it. Frank was a meticulous and very observant scientist, and he fairly quickly came to realize that truffles were never found growing independently but were always in the vicinity of oaks, filberts, and certain other forest trees. At first he suspected the truffles of being weak parasites, but eventually, through well-designed and carefully conducted experiments, he was able to prove that the below-ground hyphae of the truffles form a very important, mutually beneficial symbiosis with the roots of the trees that they inhabit. It was Frank who coined the term "mycorrhizae" to describe this fungus-root partnership, in a classic paper published in 1885. He concluded that the mycorrhiza "functions in a nutritional capacity as a wet nurse of the tree."

Although Professor Frank is given credit for the discovery of mycorrhizae, scientific historians have found that others studied other kinds of mycorrhizae before him. Frank was aware of this earlier work, and it undoubtedly laid the groundwork for his more conclusive experiments. In the mid-1800s, scientists had identified a fungus growing on the roots of small nongreen plants in the genus *Monotropa* and another type associated with orchids, both of which we now know to be mycorrhizal. Several of these researchers suspected some type of reciprocal relationship between the fungi and their plant hosts, but their experiments were inconclusive, and the subject remained controversial until Frank came along. In addition to studying truffles, Frank also spent a great deal of time studying *Monotropa* plants and their associated fungi, and this research would help him elucidate the function of mycorrhizae.

The mutually beneficial root fungi that Frank and the other nineteenth-century biologists discovered were all ectomycorrhizal. Although these were the latest to evolve, it is not surprising that they would be the first to be discovered, since the short stubby roots and thick hyphal covering they produce are visible to the naked eye. Also, their reproductive structures are quite large and obvious, many even more so than truffles, because they form as aboveground mushrooms.

It is quite impressive that within just a decade of Frank's pioneering work with ectomycorrhizae, scientists also discovered the arbuscular mycorrhizae, which can be seen only with a microscope. The latter were initially thought to be parasitic. In 1905 a French researcher named I. Gallaud published a set of outstanding drawings of his microscopic observations of this fungal-root interaction (figure 5.1), which he referred to as endomycorrhizae. These drawings were our first view of this most ancient type of mycorrhizae, and Gallaud's drawings are still used in some modern textbooks.

SOME PALEOBIOLOGISTS HAVE SET FORTH the intriguing speculation that plant roots may have evolved from the earliest fungal symbionts of rootless green algae. The evolutionary progression from an intimate symbiosis between two species to their complete integration into a single organism has been documented in other cases. The biologist Lynn Margulis, one of the early proponents of this form of evolution, argued for many years that the green chloroplast organelles (where photosynthesis takes place) found in the cells of green plants originated from an ancient symbiosis between a photosynthesizing cyanobacteria and a larger single-celled (eukaryotic) organism. Many were skeptical of Margulis's "endosymbiosis" theory, even when it was discovered that chloroplasts contain their own genetic material that is separate from the genes contained in the nuclei of the plant cells they inhabit. The matter has more or less been settled by recent comparative analyses of the nucleotide sequences of chloroplast and cyanobacteria genetic material, which reveal a remarkable similarity. Most now accept the endosymbiotic pathway for the evolution of photosynthesizing plant cells. However, no trace of genetic material to prove a fungal heritage for plant roots can be found, and few plant biologists are convinced that roots evolved from fungi.

It is only in the relatively recent evolutionary past that a handful of plant families have gained independence from the symbiosis with fungi. The exceptions to the rule, which can literally be counted on one hand, include: *Chenopodiaceae,* common species being spinach and the weed known as lambsquarters; *Brassicaceae,* such as cabbage, broccoli, and wild mustard; and *Amaranthaceae,* such as edible amaranth and pigweed. Even among these plant families, some particular species are mycorrhizal.

The next time you are walking among the mighty giants of a temperate or tropical forest, hiking through a grassy meadow, mowing your lawn, or puttering in the flower or vegetable garden, take a look around. Most of the plants that you see, if they don't fall into one of those three exceptional groups, are probably thriving through the good graces of their subsurface symbiosis with root fungi. Mycorrhizae are particularly essential for survival during those periods of water and nutrient stress that almost every plant must occasionally face—the "ecological crunch," as Michael Allen calls it. Without the fragile, gossamer-like net of subterranean fungal hyphae at their base, the towering redwoods, oaks, pines, and eucalyptus of our forests would collapse during hard times. Beneath every great tree is a fungus, you could say. And the same can be said of most other plants as well. Mycorrhizal fungi form the foundation of most terrestrial ecosystems on the planet, from our orchards, vineyards, and other farmlands to the vast savannas of Africa, the heathlands of Scotland, the tropical rain forests of South America, and the deserts of the American Southwest.

So just what is it about mycorrhizae that makes plants willing to devote as much as 20 to 30 percent of their carbon and energy to support them? What are they doing that plant roots cannot do by themselves? The secret lies in the unique structure of the fungi: Because the hyphal threads are an order of magnitude finer than the finest of root hairs, they provide access to nooks and crannies in the soil that could not otherwise be penetrated. This capacity is especially helpful to plants in acquiring certain nutrients, such as phosphorus, potassium, copper, and zinc, that do not move freely with the flow of water being taken up by roots. The finest of root hairs has a diameter of 20 to 30 micrometers (about the diameter of a hair pulled from your arm), while the diameter of a strand of mycorrhizal hyphae is only one to two micrometers.

A plant's capacity to exploit a patch of soil expands tremendously with the prolific growth of its subterranean fungal partners. If you took a cubic centimeter of soil (about a teaspoonful) from the root zone of a mycorrhizal plant and spread all of the bits and pieces of root and root hairs end to end, the total length might measure a few inches. In that same volume, the length of mycorrhizal hyphae, if completely unraveled, might range from 60 to 120 feet (20 to 40 meters)! It is primarily the superiority of mycorrhizae at mining the soil for water and nutrients that makes the association worth the cost to the plant. Some analysts have speculated that in some mycorrhizal associations the roots are doing little more for the plant than serving as a vehicle to transport the attached fungi to deeper soil layers.

In some ecosystems, the mycorrhizal fungi function not just as passive absorbers of nutrients but also as active decomposers. Like many other fungi involved in decomposition, they are capable of releasing powerful enzymes that can externally "digest" wood and other organic matter. Before the liberated nutrients have a chance to float away in the soil environment, they are immediately snapped up by the fungus and directly transmitted to the plant hosts. This short-circuiting of the nutrient cycle is particularly valuable to plants in tropical ecosystems, where heavy rains often wash free-floating soil nutrients below the root zone before they can be absorbed.

Evidence has been accumulating over the past several decades that, in addition to enhancing the function of the individual roots they inhabit, mycorrhizae also often serve as a living subterranean connection between plants of different species, through which water, nutrients, and possibly other substances can be transferred. Because mycorrhizal fungi are not nearly so host-specific as the nitrogen-fixing bacteria discussed earlier, they often spread from plant to plant and species to species. The fungi do this for purely selfish reasons, of course, not with the objective of creating a pipeline between plants. For the fungi, it is to their advantage to attach to any plant that will have them, as a means of maximizing their intake of the products of photosynthesis.

The movement of nutrients from plant to plant through mycorrhizae was first clearly demonstrated in a field experiment conducted in the mid-1960s. Researchers applied radioactively labeled calcium and phosphorus to the cut stump of a maple tree and then tracked the movement of the calcium and phosphorus into attached mycorrhizae and eventually into adjacent plants. Since then, the movement of calcium, phosphorus, carbon, and nitrogen from plant to plant has been demonstrated in many plant species and in many ecosystems.

In several studies, scientists have documented the movement of atmospheric nitrogen fixed by legume plants to adjacent nonlegume plants through a mycorrhizal conduit. Nitrogen transfer from clover and soybean to maize has been demonstrated, and in one study, as much as 15 percent of the nitrogen fixed by a species of alder tree was transferred to nearby pine trees through a fungal connection. This is an amazing phenomenon of unwitting cooperation. It requires the symbiosis between nitrogen-fixing bacteria and their legume host plant, as well as a willing mycorrhizal fungus attached simultaneously to the legume and nonlegume plant to act as the pipeline for transport of the nitrogen.

The importance of the underground plant-to-plant mycorrhizal conduits remains a matter of debate, but many are convinced that in some ecosystems the sharing of resources through such networks is so great that the plant communities function as a unified "guild" and the distinction between individual plants becomes blurred.

How large are these underground networks? No one knows for sure. It is quite possible that plants and mycorrhizae of many species are loosely linked together over tracts of land measuring many acres. We are not at all close to answering two related questions: What is the maximum distance that nutrients or other substances can be transported through mycorrhizae? And to what extent does this capacity lead to a sharing of resources among plants? What we do know is that the spread of an individual fungus organism can be quite substantial—for example, "fairy rings" of ectomycorrhizal mushrooms several meters in diameter are commonly observed surrounding pine, oak, or other host trees. A fairy ring is the aboveground manifestation of an individual subterranean fungus that may be several hundred years old. The fairy ring pattern results from the fact that these fungi grow slowly outward from a central point, and the reproductive structures (mushrooms) pop up from the perimeter, where the most active and healthiest parts of the fungus are.

Some nonmycorrhizal types of soil fungi that have been discovered are among the oldest and largest living creatures on Earth. For example, in 1992 genetic analysis of samples of the wood-eating fungus *Armillaria bulbosa*, collected in a Michigan hardwood forest over an area equivalent to several football fields, showed that it was a single organism that had been alive and remained genetically stable for more than 1,500 years. The estimated weight of this individual fungus was 220,000 pounds (100,000 kilograms)—equivalent to the weight of a blue whale! So far, no beneficial mycorrhizal fungus as large as this hefty monster has been found, but undoubtedly a series of mycorrhizae, networking from root to root and plant to plant, could encompass a very large area.

Even in the absence of physical plant-to-plant hyphal connections, the ubiquitous mycorrhizal symbiosis between individual plants and fungi plays a key role in linking the activities of subterranean creatures with life at the surface. In the broadest terms, it provides aboveground life with greater access to the water and nutrients stored in the soil, while supplying life in the underground with greater access to the carbon and solar energy collected by plants. The productivity of virtually all terrestrial ecosystems relies on this exchange of energy, water, and nutrients between the surface and subsurface.

ALTHOUGH WE HAVE KNOWN about mycorrhizae for more than one hundred years, it is only very recently that their essential role in the functioning of most terrestrial ecosystems, and in the evolution of land plants, has come to be fully appreciated. For much of the twentieth century, most scientists were very skeptical about cooperation between species. The reports of mycorrhizae were viewed as intriguing but isolated incidents that were ecologically significant only in very special environmental circumstances. When I was in graduate school in the early 1980s, a few ecologists had begun to recognize that we had underestimated the prevalence and importance of mycorrhizae, but it is only in the past ten years that the textbooks have caught up with this understanding.

For a long time, scientists had trouble reconciling the co-evolution of mutually beneficial symbioses with twentieth-century discoveries of the "selfish" mode of gene action. Also, mathematical models, which were new to ecology in the 1970s and 1980s, "proved" that mutually beneficial symbioses between two species were inherently unstable. These computer simulations led to the conclusion that successful cooperation between species would seldom persist because "cheating" is usually too advantageous—at least at the individual level and in the short term, the level at which evolutionary forces operate.

Today, less than twenty years later, there has been a complete turnaround in our thinking. This has been due in part to the sheer weight of empirical evidence that has demonstrated the essential function and ubiquity of mycorrhizal associations with plants. This research was carried out primarily during the 1970s and 1980s by a handful of dedicated soil ecologists who were not dissuaded by the majority opinion that their work with soil root fungi was esoteric and insignificant. Over and over again they showed that, as long as samples were handled correctly to preserve their integrity, wherever there were active plant roots, active mycorrhizal fungi collaborating with them were also present.

The genetic evidence—the "icing on the cake"—has been attainable only in recent years. With today's technology, we can grind up a tiny sample of root or soil and isolate the fragments of DNA released; if mycorrhizal DNA is present, it will pair up with "fingerprint" nucleotide sequences of known mycorrhizal DNA that are added to the mix. This methodology sounds complex, and it is, but today's automated genetic analysis methods are much faster than tedious microscopic examination. Most important, we can now make positive identifications even in samples in which most of the fungi have been lost or killed by mishandling and would never have been identified otherwise.

Today it is widely accepted that nearly 90 percent of plant species enter into mycorrhizal associations with fungi. This recognition of the prevalence and profound importance of this subterranean symbiosis has had an impact beyond soil ecology. Our understanding of the forces of evolution and the mathematical models we use to depict species interactions have matured as a result of this work. Evolutionary biologists now recognize that "even with selfish genes at the helm, nice guys can finish first," as Richard Dawkins, author of *The Selfish Gene,* so aptly puts it.

The mathematical population models used by ecologists have been improved to better reflect this reality. The early models were mostly equilibrium models, which assumed that nature reaches a more or less steady-state situation with regard to the population levels of species. Today we have a greater appreciation for the fact that in the real world there are simply too many natural disturbances—invasions by new species, dramatic weather events, forest fires, plagues—for equilibrium to ever be reached. And it turns out that when disturbance is commonplace, cooperation, both in our mathematical models and in the natural world, can persist and often flourish. We have also learned that the dominance of one partner over another can be prevented when the symbiosis involves a third-party species that is a predator of one of the partners. (For example, in nature, small soil arthropods such as springtails often feed on mycorrhizal fungi, perhaps keeping them in check so that they do not become pathogenic to their plant hosts.) The newer models have also been reconfigured to take into account the fact that a little "cheating" is tolerable when the partners are as distinct in size and morphology as are plants and fungi. All of these changes have led to a better match between the real world, where symbiosis clearly thrives, and the simulated world predicted by mathematical models.

6

⁓⁕⁓

WHEN THE HUMBLE
EXPLAIN THE GREAT

*Worms have played a more important part in the
history of the world than most persons would at
first suppose.*

—CHARLES DARWIN, *THE FORMATION OF VEGETABLE
MOULD THROUGH THE ACTION OF WORMS* (1881)

ONE EVENING MORE THAN A CENTURY AGO, several years prior to
Professor Frank's discovery of the symbiotic root fungi, an aging
British scientist was wrapping up his investigations of another subterranean life form—the earthworm. The scientist, in ill health, sat
crouched over a large oak desk, hastily scratching notes into a
leather-covered notebook. The study in which he worked was also his
laboratory. The desk was strewn with various dissecting tools, a
microscope, a ruler, unopened correspondence, stacks of books, and
a large earthworm munching on a cabbage leaf.

The objective of the experiment in progress was to determine the
food preferences of the common earthworm, *Lumbricus terrestris*. A
mundane topic, to be sure, but this scientist knew better than to shun
the mundane. Although a humble man, he had confidence in his ability to extract valuable insights from careful observation of the ordinary.

Glass-covered pots containing soil and more worms filled the
shelves along the walls, and pots were also tucked away in various
nooks and crannies among the room's furnishings. These experiments were the finishing touches on what had been a lifelong project.
There had been many significant distractions along the way, but he
was now optimistic that he could finish the work before his health
completely failed him.

For future contemplation he scribbled into the notebook: "Food preference = sense of taste?" He looked up for a moment and grimaced as a piercing pain rippled through his chest, and then subsided. He returned to the task at hand. He hoped to complete this series of observations before his concerned wife, Emma, came in to make him stop for the day.

From his experiments it was clear that the worms preferred green cabbage to red, celery to both of these, and raw carrots above all. Earlier he had found that they could locate buried pieces of decayed cabbage leaves and onion bulbs, "which were devoured with much relish." He had also documented in some detail the manner in which the worms searched for drying leaves in more natural settings. They almost always grasped the leaves at the tip and dragged them a few inches into their burrows, with the petiole (stem) end left protruding from the burrow entrance. They would then eat the leaves at their leisure. This plugging of the burrows was one of the worms' strongest instincts and apparently served other functions, perhaps protecting them from predators or rapid water infiltration during heavy rains. The scientist had become fascinated by the efficiency with which they gathered the leaves, and he even suspected it revealed some level of intelligence. He had cut up paper into varying triangular shapes and found that the worms quickly "learned" how best to manipulate these artificial leaves into their burrows.

Suddenly, his thoughts were interrupted by a knock at the door. It was Emma.

"Our guests have arrived," she said in a tone that also meant, he knew, that it would have to be the end of the day's work. He had completely forgotten about the visitors for dinner.

He slowly turned and rose as two distinguished gentlemen entered the room. Despite his age and health problems, the scientist had a towering frame and quiet strength. His size, combined with his long white hair, enormous white beard, and penetrating blue eyes, gave him a commanding presence.

The two guests were famous physicians and, more important, political reformers of their day. Despite their prestigious credentials, they both appeared slightly awestruck by their face-to-face encounter with this gentle old fellow. The nervousness and intimidation showed in their faces.

FIGURE 6.1   *Charles Darwin in his later years, when he was completing his research on earthworms. Courtesy of the Wellcome Trust, London.*

Emma broke the awkward silence. "Gentlemen, let me introduce you to my husband, Mr. Charles Darwin."

Timidly, they stepped forward one by one, bowed, and shook Darwin's hand.

IT WOULD NOT BE HYPERBOLE to say that few other individuals in the history of the human race have had as great an impact on our perception of the world around us, and our place in it, as Charles

Darwin. The significance of his work was recognized, to some degree, within his lifetime. So why did a scientist of Darwin's stature spend his final years focused on a topic—the behavior of the lowly earthworm—that many would consider completely inconsequential? Was it an eccentricity of old age? Or was it perhaps a refuge from the storm of controversy he had endured for years following the publication of *On the Origin of Species* and *The Descent of Man*?

The fact is that Darwin did not consider earthworms inconsequential, and his interest in them spanned almost his entire professional career. Historians trace the beginning of this interest to the year 1837, not long after his return from the now-famous five-year *Beagle* voyage. The young Darwin had not yet written *Origin*, nor had he completely formulated the concepts of evolution and natural selection. He had gone to rest for a few weeks at the country home of his uncle, Josiah Wedgood, in Staffordshire. While they strolled one day through an unused part of the garden, his uncle Jos pointed out that lime and cinders had been spread on the surface several years before but now were buried by a few inches of soil that had been brought to the surface through the "castings," or egested organic matter and soil, of earthworms. His uncle did not expect this bit of gardening trivia to be of much interest to his nephew, who was usually concerned with scientific matters on the continental scale. But much of Darwin's early training and interest was in geology, and he was astounded by the magnitude of the effect of the earthworms, given enough time, on the landscape. The published literature on earthworms during Darwin's era only discussed these creatures as a potential minor nuisance for gardeners and farmers.

On November 1, 1837, Darwin presented to the Geological Society in London a summary of his initial observations on the soil formation activities of earthworms. It was one of his first presentations to this esteemed group. The paper also appeared in the *Proceedings and Transactions* of the society. More than forty years later, in 1881, the great naturalist would complete his earthworm research and publish the results in his final book, *The Formation of Vegetable Mould Through the Action of Worms*. The book sold phenomenally, thousands of copies within the first few weeks, presumably due more to the reputation of the author than to the appeal of the subject matter. The periodical *Punch's Almanac* poked some good fun at the celebrity scientist with an 1882 cartoon entitled "Man Is But a Worm." The

FIGURE 6.2 *Caricature from Punch, December 6, 1881, poking fun at Darwin's interest in earthworms.*

illustration shows a gradual evolution from worm, to monkey, to man, to the caricatured old Darwin himself.

Although hardly popular reading among the general public today, most modern ecologists and soil scientists are at least aware of Darwin's final book, for it contributed to the development of their disciplines. Darwin's treatise on worms is recognized as one of the first instances in which the influence of biological organisms on their physical environment was clearly documented. In *Worms*, Darwin held to a theme also reflected in his writings on evolution and geology: Small, almost imperceptible quantitative changes can accumulate over time and eventually change our world in a significant way. The key that unlocked the door to Mother Nature's secrets for Darwin, and the key to his genius, was the ability to stretch his imagination to encompass geological time—thousands of years, thousands of centuries.

THE CENTENARY OF THE PUBLICATION of Darwin's final book was celebrated in 1981 by an international symposium on earthworm ecology. The meeting, held in England, was attended by more than 150 "adrenalin-charged research workers in the full heat of peer-group interaction," according to J. E. Satchell, the symposium organizer and editor of the proceedings. Some of the papers presented at the meeting dealt with issues that Darwin had not considered, such as the potential use of earthworms in land reclamation and waste management. A significant number of the papers, however, essentially corroborated Darwin's earlier findings, utilizing more sophisticated scientific techniques. Darwin's work has stood the test of time because he had the patience to gather irrefutable evidence before reaching conclusions, and he scrupulously relied on the facts, not personal bias, in forming his theories. All scientists strive, of course, to bring this kind of patience and honesty to their work. Darwin just did it better than most.

It is important to realize that before the appearance of Darwin's book, few people recognized any beneficial attributes of earthworms. When the book was first published, a review in the farmer periodical *Country Gentleman*, in its February 2, 1881 issue, rejected Darwin's ideas on the grounds that it was common knowledge that earthworms damage plants grown in flowerpots.

Today, in contrast, earthworms are almost revered by gardeners and farmers, who view them as icons of a healthy, productive soil. This perception permeates our culture well beyond the bounds of those directly involved with crop production or land management. The humble earthworm has been put on a pedestal that we reserve for only the most valued and esteemed of fellow creatures.

Young children are usually fascinated by the creepy-crawly nature of worms, and also seem to know quite a lot about them. Information about the behavioral and physiological peculiarities of worms and an appreciation for their role as caretakers of the soil are handed down from generation to generation, like folklore.

One bit of earthworm trivia many of us learn at an early age is that individual worms are both male and female (hermaphroditic). If you have grown up with the mistaken impression that because of their hermaphroditic nature worms do not mate with a partner and therefore have a rather dull sex life, let me set the record straight. Most worms are sexually active year-round and routinely copulate with others of their kind. Some species are capable of reproducing asexually, it is true, but this is the exception rather than the rule. Most worms, including the common *L. terrestris,* can breed only sexually. A typical "sexual embrace" can last the better part of an hour. Since they fully enjoy the sexual experience as both a male and a female simultaneously during the sexual encounter (imagine that!), it's no wonder they don't rush things. Darwin didn't dwell on this aspect of earthworm biology in his book. It was, after all, the Victorian era. He did note, however, that "their sexual passion is strong enough to overcome for a time their dread of light."

During sex, earthworms lie with their stomachs together and heads pointed in opposite directions, X-rated-movie style, in order to align the male pores with the female clitellar region. Both the male and female parts are on segments near the middle of the body. The clitellum is easily recognized on most worms as a slightly swollen area. It is essential that copulating worms stay close together for the exchange of seminal fluids. To accomplish this, the worms flex inward and grip each other by piercing their mate with the small pointed and grooved hairs, or setea, that project from segments near the clitellum. Large quantities of mucus are secreted, and they may clasp and release each other several times during a single sexual encounter.

After copulation, each worm secretes a special fluid that surrounds its clitellar region and eventually hardens over the outer surface. The worm then squirms backward, drawing the hardened tube that was surrounding the clitellum over its head (figure 6.3). When the worm is completely free, the ends of the hardened tube close together, forming a lemon-shaped cocoon. The cocoon contains the ova and spermatozoa, as well as a nutritive fluid produced by clitellar gland cells. Fertilization actually occurs in the cocoon, after the sexual act and external to the parent worms. A typical cocoon may contain one to twenty hatchlings, although seldom more than one or two survive. Depending on the species and environmental conditions, it can take a few weeks to five months for them to hatch.

One would expect that this kind of sexual activity would call for some heavy breathing, but that's another funny thing about earthworms—they have no lungs and don't really breathe like most other animals. The oxygen they need must diffuse directly through their skin and into their blood, which contains hemoglobin, a respiratory pigment also found in the blood of humans and other animals. Earthworm hemoglobin is an improved version of human hemoglobin in that it has a higher affinity for oxygen, compensating to some extent for the inefficiency of a respiration system dependent entirely on passive diffusion. Embedded in the body wall of earthworms are highly branched capillary blood vessels that maximize exposure of the blood supply to the outer atmosphere.

In all respiratory systems, oxygen must first be dissolved in an aqueous (water) layer on the respiratory surface before it can be absorbed by the blood. Most animals have evolved to cope with this requirement by having their respiratory surface inside the body, protected from harsh drying air. However, in earthworms the entire external surface of the body serves as lungs. It is essential that this surface be kept moist at all times by secretions from epidermal mucous glands. Dry environmental conditions can spell death for a worm. Thus, burrowing species burrow more deeply when upper soil layers become dry, and though earthworms are found in virtually all terrestrial ecosystems on the planet, they're conspicuously absent from our dry deserts and arctic regions.

Because Darwin focused on the role of the earthworm in organic matter decomposition and soil formation, he studied its digestive system and feeding behavior more than its respiratory system. The issue of food selection and preferences led him to consider whether earthworms could actually taste or smell food. This research eventually expanded to include experiments designed to reveal other sensory capabilities of the earthworm, such as their ability to see, hear, and feel.

During this phase of his investigations, on sleepless nights, Darwin would fumble around in the study in the dark, shining light from various sources, and of various colors, at the worms to evaluate their sense of sight and response to light. He concluded (correctly) that they are essentially blind but do have some means of sensing light

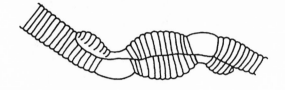

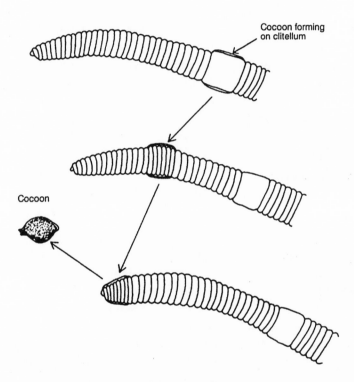

FIGURE 6.3 *Sexual mating of earthworm pair and release of the cocoon. From C. A. Edwards and P. J. Bohlen,* Biology and Ecology of Earthworms, *3d ed. (London: Chapman and Hall, 1996).*

intensity, particularly at their anterior (head) end. They are thus able to distinguish between night and day and to avoid daytime predators. This ability explains the nocturnal nature of the common earthworm, *L. terrestris* (also know as "nightcrawler").

To determine whether his worms could hear, Darwin got the entire family involved. He noted that the worms "took not the least notice of the shrill notes of a metal whistle" (blown by his grandson Bernard), "nor did they of the deepest and loudest tones of a bassoon" (played by his son Frank), and when they were placed near to the piano played as loudly as possible by Emma, "they remained perfectly quiet." He concluded (again correctly) that earthworms are deaf, but he also recognized that extreme tactile sensitivity compensated somewhat for their deafness. When the pots containing worms were placed directly on top of the piano and Darwin himself banged out one note at a time, the worms "instantly retreated into their burrows" in response to the vibrations. Darwin wrote, a bit anthropomorphically by today's standards: "It may be well to remember how perfect the sense of touch becomes in a man when born blind and deaf, as are worms."

DARWIN AND MOST EARTHWORM researchers who continued after him have focused their work on a handful of species, most in the family *Lumbricidae.* There is little doubt that the lumbricids are ecologically the most important family group in the Northern Hemisphere. Members of another family, the *Megascolecidae,* are equally important in the Southern Hemisphere. Worldwide there are at least twelve other taxonomically distinct earthworm families and more than twelve hundred species. Unfortunately, there is very little to say about most of these because they have not been well studied.

The lumbricids are a spectacularly successful family, and so it is not surprising that they have attracted most of the attention. They tend to quickly dominate other species when introduced into a new environment, and they are widely adapted. Many American fishermen might be surprised to learn that their favorite bait species, the seemingly ubiquitous nightcrawler, is not native to North America but was introduced, along with many other lumbricid species, by European settlers.

One reason for the success of *L. terrestris* and some other burrowing worms is that they can migrate to deeper soil when conditions near the surface become unfavorable. They can also enter a hibernation-like state and remain coiled up deep in their burrows, often at a depth of three to six feet (one to two meters), to weather out an extended bad situation. *L. terrestris* worms have been reported to live for four to six years or longer when kept out of harm's way in a controlled environment. In the field, it is assumed that only the very lucky or very clever make it to this old and senile stage. Most probably die in some tragic fashion within a year or two of leaving the cocoon.

Lumbricids, like most other worms, can tolerate wet and cold conditions much better than hot and dry. The soil can become too wet after heavy rains, however, and the worms may come to the surface in mass migrations in order to find sufficient oxygen. Most of us have witnessed at one time or another one of these mass migrations. It isn't always a successful survival strategy. Many fatalities occur upon exposure to drying air and ultraviolet radiation.

Darwin did not spend so many years contemplating his lumbricid worms out of idle curiosity. It was his suspicion that they played an important, and previously unrecognized, role in soil formation that drew his interest. His approach was primarily that of a geologist, or perhaps it would be more appropriate to say he came to this study as an ecologist, even though ecology had not yet been defined as a scientific discipline.

To prove his primary thesis about the impact of earthworms on their physical surroundings, Darwin conducted a long-term experiment that revealed a patience most scientists in today's "publish or perish" environment can only envy. He carefully spread a uniform layer of chalk over a section of grassy field in 1842 and returned in 1871, twenty-nine years later, to dig a trench and measure the depth to which the chalk had been buried by soil and "vegetable mould" (humus) brought to the surface by worms. The measured depth of new soil was more than six inches, or 0.22 inches per year. This measurement corroborated more circumstantial evidence he had gathered from various sources over his decades of research. He wrote: "I was thus led to conclude that all the vegetable mould over the whole country has passed many times through, and will again pass many times through, the intestinal canals of worms. . . . It may be doubted whether there are many other animals which have played so important a part in the history of the world, as have these lowly organised creatures." Investigations of the past century have provided overwhelming evidence in support of these claims.

Darwin was intrigued by the importance of the earthworm soil formation activity in the burial and preservation of archaeological ruins, and he dedicated a chapter of *Worms* to this subject. In his later years, despite his poor health and the admonitions of his wife, he made several trips to Stonehenge. There he spent hours digging in the hot sun in order to document the role that worms had played in burying, or partially burying, many of the prostrate druidical monoliths at the ancient historic site.

It is now widely recognized that earthworms are in some ways the terrestrial equivalent of the filter-feeders of the sea, consuming 10 to 30 percent of their live body weight per day. In some ecosystems, they may consume nearly 100 percent of total leaf fall during the course of a year. Earthworms essentially act as biological blenders, fragmenting plant debris, mixing it with soil and living and dead microbial biomass, and exposing organic surface area for further transformation into humus by the microbial decomposers. When earthworms are not present or are not active because of environmental conditions, thick mats of dried leaf material accumulate, and soil quality deteriorates. In some tropical ecosystems, termites fill the important ecological niche occupied by earthworms. Ants also can play a similar role to some extent, but they are no match for earthworms when it comes to quantity of plant debris consumed and soil moved about.

The partnership between earthworms and microbes is essential to the success of both groups of organisms. The casts excreted by earthworms are heavily colonized by microbes because they are rich in nutrients. Conversely, earthworms do not live on plant debris alone; they need microbes in their diet to meet their nutritional needs. In one experiment, it was found that worms do not reach sexual maturity if protozoans are excluded from their food supply.

Many of the microbes consumed by earthworms are not actually digested but end up being transported great distances (from the perspective of a microbe) before they are literally "cast out," in a completely viable state, at a new location. Both beneficial and pathogenic microbes are dispersed in this way. It has been demonstrated that the presence of earthworms at a site can substantially increase nodulation of legume plants by *Rhizobium* nitrogen-fixing bacteria. In one study, more than half of the earthworms examined contained beneficial mycorrhizal root fungi in their guts, and these fungi could survive for twelve months in worm casts. Although earthworms can spread soilborne disease from plant to plant, they also can spread the natural enemies of plant and animal pathogens. Earthworms were observed to significantly *reduce* the spread of harmful apple scab fungi in an apple orchard by pulling infected fallen leaves containing spores into their burrows, thus reducing the number of spores at the surface.

As earthworms excrete their casts above- and below-ground, microbial communities and nutrients become vertically mixed within and below the plant root zone. In the long term, this mixing affects the depth and composition of the upper soil layers, or "horizons." The volume of an individual cast produced by most common European and North American earthworms would only fill a tablespoon (about ten grams), but on an acre basis and over the course of an entire year, twenty to thirty tons of soil is typically moved from below- to aboveground by earthworm activity. Some studies have estimated the amount of soil moved at one hundred tons per acre, equivalent to nearly a half-inch depth of soil (over one centimeter). This soil mixing benefits farmers and gardeners but is a nuisance for managers of golf courses and tennis greens, who sometimes use pesticides to control earthworms to avoid the deposit of casts at the surface. Some large exotic earthworms species in Africa and Asia can be a real nightmare for lawn turf managers because they produce towerlike casts several inches in height (ten to twenty centimeters) and weighing more than three pounds (one and a half kilograms). Darwin was so impressed by the giant casts sent to him by colleagues in India that he included engravings of some in his book (figure 6.4).

Earthworms have a very positive effect on soil structure, or "tilth," and consequently on plant growth. Earthworm burrows create soil biopores that allow water and air penetration deeper into the soil profile, benefiting both soil microbes and plant roots. Some plant roots, particularly in compacted soils, follow the path of abandoned earthworm burrows and thus are able to explore a greater portion of the soil for nutrients and water. Earthworms also improve soil tilth by secreting gummy sugar compounds that help hold particles of soil together into small aggregates. As a result, the soil develops a desirable, crumblike structure that improves water infiltration and water-holding capacity compared to soils with few worms.

Although the importance of earthworms for crop production would dominate any assessment of their economic impact, they are of value for many other reasons as well. Fishermen would immedi-

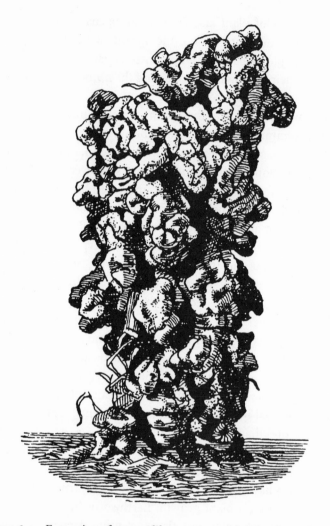

FIGURE 6.4 *Engraving of a towerlike cast (actual height about three inches) produced by a large earthworm species and submitted to Darwin by a colleague at the Botanic Garden of Calcutta. From Charles Darwin, The Formation of Vegetable Mould Through the Action of Worms, with Observations of Their Habits (London: Murray, 1881).*

ately think of the fish bait market, which is indeed big business in some regions. The value of Canada's bait industry is estimated to be more than $50 million per year. Throughout an important worm-harvesting area near Toronto, Canada, worm-picking crews scour golf courses and pastures in the middle of the night from April through October to capture Canadian nightcrawlers. They use miners' lamps for illumination and strap an open tin can to each ankle. One of the cans contains sawdust that the pickers dip their fingers into to help them grip the slippery worms, and the other, larger can is used to hold up to five hundred worms. A good picker can find ten thousand worms on a good night. Attempts to raise nightcrawlers in compost bins and the like have proven difficult and not as economical as this harvesting approach. Other earthworm species are much easier to raise "in captivity" than *L. terrestris,* but owing to their smaller size, among other reasons, they are considered inferior as bait.

Earthworms are increasingly being evaluated for their use in land reclamation and organic waste management. Many outstanding successes have been reported, but also some failures. The failures are usually due to a poor matching of earthworm species to the task and environment at hand, and to poor management. For example, reclamation of former mining sites often requires special species that can tolerate acid soils. In many cases, the earthworms must be supplemented with nutrients lacking in the soil environment in order to become established.

Nevertheless, the research efforts in this area will undoubtedly lead to many new practical uses for earthworms in the not-too-distant future.

DURING HIS LATER YEARS, as he was trying to complete his book on worms, Darwin was constantly distracted by the fact that his theory of evolution had become a political issue. Out of his control, the theory had become linked with political movements for a lessening of the power of the Church of England, for land nationalization, and for workers' rights. Darwin had difficulty relating to many of the zealots who besieged him for his support. His aristocratic roots made the concept of land reform difficult to accept. In addition, many of the people behind these movements were atheists, and while the basically agnostic Darwin could accept this, his wife Emma, whom he loved dearly, was a devout Christian and could not. Finally, Darwin was a scientist, not a politician.

In one awkward dinner at the Darwin home, Charles was seated with Emma, the local parson, and two of the more famous and distinguished atheist political reformers of the day, one British and the other German. The reformers were genuinely horrified to learn that the author of the *Origin,* who had been elevated to heroic status by their movement and who they hoped would become a noble ally, was spending his days groveling in the dirt with worms. It came up during the dinner's first course. Why had he stooped to a subject so insignificant? they wanted to know. It is reported that Darwin turned gravely toward them at this moment and said simply, "I have been studying their habits for forty years." Not surprisingly, it was difficult for the politically ambitious guests to see any link or similarity between Darwin's interests in geology and evolution and his work with his beloved worms.

Darwin, like the creatures that were the subject of his final book, often worked at a painstakingly slow pace. He proceeded with determination, sincerity, and a humility born of his appreciation for the awesome power of nature's slowly churning forces, which he so wanted to understand. Step by step, he developed the necessary evidence for his theories that would shake the world. His scientific achievements are incredibly broad in scope, but there is at least one central theme: Small changes, through unimaginable time, can cause profound consequences. This is how the lowly worm, in partnership with other soil-dwellers, forms the earth we tread on, farm on, build our skyscrapers on. This is how the humble can explain the great.

7

GERM WARFARE

If one considers the period for which animals and plants have existed on this planet and the great numbers of disease-producing microbes that must have gained entrance into the soil, one can only wonder that the soil harbors so few bacteria capable of causing infectious diseases in man and in animals.

—SELMAN WAKSMAN (1940)

The Lord created medicines out of the Earth, and he that is wise will not despise them.

—ECCLESIASTICUS 38:4

THE SUBTERRANEAN IS A PLACE WITHOUT DISCRIMINATIONS, where all that we discard in the hustle and bustle of living and dying is taken in without resentment. It is a place where waste has no meaning. As we go about our oh-so-important day-to-day surface activities, the thriving community of organisms that inhabit the underground is hard at work, not only recycling and supplying plants and animals with essential nutrients, but also challenging the pathogens and neutralizing the toxins we so thoughtlessly dump into our environment.

As we shall see, the natural "checks and balances" within the soil microbial community normally keep the number of disease-causing organisms to a minimum. However, Mother was right: You should wash your hands after playing in the dirt, especially if you have an open wound. Although most common wound infections are caused by bacteria that live on our skin (*Staphylococcus aureus* and *Streptococcus* species), there are a few microbial villains lurking in the underground that you don't want to have to do battle with.

Historically, the most serious threat to human health of all soil-borne pathogens has undoubtedly been *Clostridium tetani,* the causative agent of tetanus. This pathogen has been feared by humans for millennia. The first official record of tetanus is found in the Egyptian hieroglyphics of the Edwin Smith Papyrus, dated 1500 B.C., where a patient with a penetrating scalp wound and suffering from "lockjaw" is described. Hippocrates, in 400 B.C., provided the gory details of several fatal cases.

Although we now have a vaccine to prevent tetanus, we have yet to develop a cure for nonvaccinated individuals who become infected. And there are a surprising number of such people in some developing nations. Fatality rates can be as high as 70 percent for victims of tetanus in remote areas where medical treatment is limited. There are antibiotics available to kill the bacteria, but there is no effective antidote for the lethal toxin that the bacteria release once they reach the nerve-muscle junctions in the body. This neurotoxin, called tetanospasmin, causes excruciatingly painful symptoms in human victims. It leads to uncontrolled spasms of skeletal muscles and severe cramping. The tension and cramping usually begin around the wound as well as the jaw muscles, causing trismus, the medical term for lockjaw. Unfortunately, patients often don't arrive at hospital emergency rooms until they are at this stage. By then, doctors can only try to manage the disease with heavy doses of antibiotics and drugs to neutralize any toxin that has not already reached the nervous system.

Within just a few days, tetanus symptoms can proceed to the back, legs, and arms, where muscle contractions can become so severe that muscle tearing and compression fractures of the vertebrae occur. A horrifying condition known as opisthotonos sometimes sets in: The spine bows backward so severely that the heels and back approach each other (figure 7.1). Death can result within just a week or two, usually due to spasms of the diaphragm and other respiratory muscles, or cardiac arrest. For cases this severe, patients are given powerful muscle relaxants that essentially paralyze them, and they must be connected to a respirator until the disease runs its course. Although deep puncture wounds are most likely to cause tetanus, patients have been known to die from wounds as minor as a splinter.

Pick up a handful of soil just about anywhere in the world and you will be holding spores of the tetanus bacterium. An individual spore

FIGURE 7.1 *Painting by Charles Bell (1809) of a soldier dying of tetanus during the Napoleonic Wars. Note the extreme contraction of the trunk and leg muscles, causing the characteristic opisthotonos—backward bowing of the back. Courtesy of the Royal College of Surgeons of Edinburgh.*

can remain viable for as long as forty years. Like most other *Clostridium* species, *C. tetani* is anaerobic. When these spores find themselves in the low-oxygen environment of a deep wound (or within a minor wound that has closed up), they germinate and the bacteria begin to reproduce. As time goes on and some members of the growing bacterial population die, they release their potent neurotoxin.

It was not until the late nineteenth century that the causative bacteria was identified; a vaccine to prevent tetanus was then developed in a collaborative effort between Emil von Behring of Germany and Shibasaburo Kitasato of Japan. Both men had been students of the famous microbiologist Robert Koch in Germany. By about 1920, the vaccine was widely available in the United States, Europe, and Australia.

Massive immunization programs have all but eliminated tetanus in the developed world, although the same cannot be said for parts of

Africa and Asia. Where programs are in place, the first vaccination is administered in infancy, with follow-up boosters during the teen years and early adulthood. Thereafter, a booster every ten years is recommended to maintain immunity. In the United States, only about a hundred cases are reported per year—fewer than the number of new cases of leprosy reported annually. Most tetanus victims in this country are intravenous drug users who have not kept up with boosters, or very old patients who have not had a booster for several decades.

In shocking contrast to the control achieved in developed nations, tetanus remains the second leading cause of infant mortality worldwide among diseases for which there is an effective vaccine. Most infant deaths are caused by infection of the umbilical stump due to nonsterile techniques during delivery and lack of immunization of the mothers. In 1997 an estimated 250,000 babies lost their lives to tetanus—88,000 in Southeast Asia and 95,000 in Africa. So many children meeting such a cruel fate within days of their entry into the world, and it all could be prevented! Since 1989, the United Nations World Health Organization (WHO) has focused on improving this disgraceful situation, with considerable success. As disturbing as the 1997 statistics are, they reflect a significant improvement from 1990, and health officials are optimistic about continued reduction in tetanus fatalities in the future.

There are a few other bacterial species within the genus *Clostridium* that can do serious damage to humans if they gain entry into the low-oxygen environment of a wound and germinate. The condition known as "gas gangrene" can be caused by *C. perfringen, C. novyi,* or *C. septicum,* all of which are common in many soils. Our body has natural defenses to fend off these bacteria, but if they get the upper hand, hang on! You could be in for the fight of your life. Once established, they rapidly spread from the original point of infection, destroying living tissue in their wake as they spread through limbs and other body parts. Control in advanced cases can be difficult: Sometimes dead, infected tissue must be surgically removed or limbs must be amputated. Gas gangrene is most often a threat when medical treatment of a serious injury is delayed, after a car accident, for instance, or in military combat.

Soil fungal species can also pose a health risk to humans, but only a couple of these are endemic to the United States. The worst of these is called blastomycosis, caused by the fungus *Blastomyces dermati-*

tidis. It is found predominantly in the soils of the Mississippi and Ohio river basins. What is spooky about this particular soil germ is that, unlike most, it does not require an open wound to enter the human body. The initial infection commonly begins with inhalation of dust containing spores of the fungus. It may stay in the lungs, but often spreads to the skin, where ulcers and abscesses form. It can be successfully controlled by drugs, but surgery may be necessary for drainage of large abscesses, and thirty to sixty deaths are reported each year in the United States.

The second soilborne fungus of concern in the United States, *Coccidioides immitis,* causes "valley fever." This species lives in the dry, highly alkaline soils of the San Joaquin Valley in central California and other semidesert habitats throughout the Americas. Like blastomycosis, this disease often infects humans when they inhale the spores. The consequences of valley fever are usually much less serious, however, than those associated with blastomycosis. In fact, most cases go unnoticed because the symptoms resemble those of a cold or the flu. Victims usually recover within two to three weeks and acquire lasting immunity, never knowing that they were coping with a fungal infection caught from the soil. It is only every now and then, usually in patients with weakened immune systems, that one of these infections becomes more serious and results in a progressive, chronic lung disease.

In the tropics, there are a number of soil fungi that can enter through the skin and cause serious, long-term infections. Barefooted agricultural workers are frequent victims of these pathogens, which usually gain entry through minor cuts and scratches on the feet. The symptoms vary with the particular pathogen and sometimes take years to develop. The disfigurements (pigmented nodules, swellings, open festering abscesses) usually show up near the site of infection and come in many shapes and sizes depending on the particular fungus species involved. Although these tropical fungal diseases cause unsightly disfigurement and often serious discomfort, most are not life-threatening.

PLANTS HAVE MANY MORE LIFE-THREATENING enemies in the soil than we do. Their roots are constantly exposed, and over the millennia quite a few soil microbes have evolved the capacity to circumvent the natural defenses of the root system of specific plant hosts. These pathogens may parasitically drain the plant of sugars and other vital fluids, release toxins that cause stunted, abnormal root growth, or clog the vessels that transport water and nutrients within the plant. If a soilborne plant pathogen gets the upper hand on its victim, it can eventually make mush of the root system, cause yellowing and wilting of the leaves, and kill its host. Food crops are often among the victims of plant pathogens, and indirectly, so are the farmers who depend on healthy crops for their livelihood.

Many of the most serious soilborne plant pathogens are fungi or fungus-like water molds. Fungal species in the genus *Fusarium* or *Verticillium* cause many of the diseases known as "wilts" or "yellows" by farmers and home gardeners. Many of the diseases that cause "root rot" and "damping off" (the collapse of young seedlings) can be blamed on one of the *Phytopthora, Pythium,* or *Rhizoctonia* pathogenic species. Fungi often can persist in the soil for one to a few years. Farmers need to rotate crops and plant nonsusceptible crops or varieties in infected fields for a year or two to prevent an escalation of disease pressure. Crop rotation is not always a feasible control option, however, as in the case of *Plasmodiophora brassicae,* the causative agent of the deadly "clubroot" fungus that attacks cabbage and related crops and can persist in the soil for seven years or more.

In addition to the fungi and water molds, several microscopic, wormlike nematodes can set up shop in plant roots and drain the plants of sugars and essential nutrients. Most of the parasitic species of nematodes are members of the genus *Pratylenchus* or *Meloidogyne.* Bacteria in the soil can also cause some root diseases, although most bacterial pathogens attack the foliage of plants.

The soil often serves as a temporary reservoir for dormant stages of fungi or bacteria that do their real damage to the aboveground parts of plants. Some of these pathogens cannot survive in the soil *per se* but can survive below-ground if remnants of stems, roots, or leaves of their host plants are buried in the soil. These latter types are often relatively easy to control by crop rotation because the plant residues they require decompose in the soil over the course of a few weeks or months.

Until recently, one of the potentially most serious plant pathogens, the fungus-like *Phytopthora infestans,* which causes potato late blight, was among those that fortunately could not survive for long in the soil. As long as farmers practiced some crop rotation and did not use infected potatoes from the previous year as seed potatoes, they could begin each growing season with a clean slate and no worries about long-term carryover of this dreaded disease in the soil. When it gets loose aboveground, *P. infestans* spreads like wildfire from plant to plant, leaving behind a wasteland of leafless, melted, blackened stems with a disgusting odor characteristic of the disease. Much of Ireland reeked of this plant disease for several summers during the 1840s, the years of the great potato famine, when more than one million Irish citizens lost their lives and many more left their homeland to escape the devastation.

The rapid spread of late blight aboveground is due to the release of hundreds of thousands of asexual reproductive structures called sporangia from every small, cottony site of infection on the leaves. These sporangia, however, usually do not survive in the soil for more than a few weeks. Because of this weakness in the survival ability of *P. infestans,* potato late blight had been brought under reasonable control by the mid-twentieth century through the use of only certified disease-free seed potatoes to prevent year-to-year carryover of the disease, and chemical fungicides to slow any aboveground outbreaks during the growing season.

All of this began to change in 1976, however, when an exotic strain of *P. infestans* was inadvertently imported into Europe (and eventually other parts of the world) from the region where the soil pathogen is native—the central highlands of Mexico. Before this migration, most of the *P. infestans* fungi throughout the world were quite similar genetically because they were all of a single mating type, called A1, and thus had been able to reproduce only asexually. The exotic Mexican strain was of the A2 mating type. Its appearance in Europe was equivalent to bringing males and females of an animal species together—sex happened. By the 1980s, both mating types of *P. infestans*—a necessity for sexual reproduction—were being discovered in all parts of the world where potatoes were grown, including the United States.

Giving *P. infestans* a sex life has had two frightening repercussions. First, sexual reproduction results in the release of dormant structures called oospores, which, unlike sporangia, can survive for years in the soil. Second, the now-rampant sexual reproduction of *P. infestans* is creating new genetic diversity, and as a result there are many new genotypes that are resistant to fungicides.

As we begin a new century, potato late blight, a disease we thought we had nearly conquered, is back, and with a vengeance. Many scientists feel that it has once again become the number-one crop disease threat to our food supply. Frustratingly, it is a problem we brought on ourselves. We should have known better than to ship potatoes into Europe from the region in Mexico where the pathogen originated and therefore has the greatest genetic diversity. Mexico is not normally a potato exporter, but in that fateful year, 1976, there was a severe seed potato shortage in Europe because of drought. Ironically, the initial introduction of late blight to Europe around 1840, and the subsequent Irish potato famine, were due to a transfer of infected plants or potatoes, probably by a scientist or amateur botanist, from the same region of Mexico.

With our latest blunder, we have turned late blight into a sexually reproducing disease that can now survive in the soil for years at a time. Rotation is a less effective control option, and the pathogen is rapidly becoming resistant to fungicides. Within just the past few years, right in my own backyard in upstate New York, several potato farmers have nearly gone out of business because of the resurgence of this disease. Globally, it is a disaster waiting to happen. Ireland is less dependent on potatoes than it was in the 1840s, but in the past century potatoes have become one of the four major world food crops (along with rice, wheat, and corn). Poor developing countries produce more than one-third of the world potato crop, and several struggling economies of central Europe and Russia are particularly at risk.

Many scientists are attempting to solve the late blight problem, but in the meantime fungicides are becoming less effective, and crop failures more common, each year. We can only hope that, with appropriate investments in research on the epidemiology and genetics of *P. infestans,* scientists will arrive at a solution before we witness another potato famine crisis as bad or worse than the famine suffered in Ireland more than a century ago.

BEFORE YOU RUSH OFF to soak the potatoes in your garden with fungicides or scrub yourself down in antibacterial soap, let me try to bring all this talk about disease back into perspective. First of all, the pathogen lifestyle among microbes in the soil, as with microbes in general, is extremely rare. Most pathogens only "go bad" and attack us or our crops under special environmental conditions and when the natural defenses of a potential victim are down. We can also be reassured by the fact that the handful of bacterial and fungal species in the soil that are potentially harmful to us or our crops have many natural enemies of their own. The pathogenic species are significantly outnumbered by thousands of others that either are above them in the subterranean food chain or compete with them for resources. Because of this, *most soils actively suppress disease.* That is, unless man steps in and destroys their natural disease-suppressive capacity by poor management or the application of broad-spectrum pesticides that kill as many beneficial creatures of the underground as harmful ones.

In the past, farmers who had chronic problems with soilborne crop diseases would sometimes, out of desperation, attempt to "sterilize" the soil by fumigating it with powerful, volatile toxins that killed almost everything. This strategy can result in short-term yield increases, but these increases may be due as much to the nutrients released by dead soil organisms as to disease control. In any case, the pathogens always come back, and when they do, there are fewer of their natural enemies in the soil to control them. This approach leads to chronic dependence on pesticides and a toxic environment that directly threatens human health. By the end of the twentieth century, farmers and agricultural scientists were beginning to realize that creatures of the underground are too numerous, too mobile, too robust, and too diverse to ever be dominated by us in this way.

There has been a major shift in recent years in how agricultural scientists and progressive farmers think about soil management. More and more of them recognize that a healthy soil need not be completely free of all potential disease organisms as long as soils are managed so that their disease-suppressive capacity is encouraged. This new approach is one of working with nature as opposed to fighting it. As we enter the twenty-first century, the practice of soil fumigation is becoming much less common and more cost-effective, environmentally sound alternatives are being developed. We have called a truce in our futile battle to completely dominate subterranean life.

The exploration of new methods to promote the disease-suppressive capacity of farm soils is one of the most exciting and challenging areas of agricultural research. Some of the methods coming out of this research are new twists on old ideas. More sophisticated approaches to traditional crop rotation are being used. For example, we are taking advantage of recent discoveries that some crops release toxins from their roots that target certain pathogens. Sudangrass is one such rotation crop that I have worked with in my own research program at Cornell. Dr. George Abawi, a plant pathologist with whom I collaborate, has found that sudangrass releases compounds from its roots that reduce the levels of pathogenic nematodes in the soil. When beans are grown in rotation after sudangrass, nematode infection is significantly reduced and bean yields are increased.

The prescription for improving the disease-suppressive capacity of soils can vary considerably depending on the particular disease, crop, and location involved. Crop rotation is not always the answer and in fact can sometimes prevent the establishment of a stable population of beneficial microbes. For example, in some wheat-growing areas it has been found that continuous wheat production on the same land for a few years, rather than rotation with other crops, is required for the populations of beneficial microbes to build to the levels necessary to control specific wheat diseases.

Compost (degraded organic matter) has been used as a soil amendment by farmers and home gardeners for centuries as a means of improving fertility and water-holding capacity, and recent research suggests that some composts also aid in disease control. Some of these composts "seed" the soil with natural enemies of pathogens, while others simply create a soil environment that promotes the growth of the beneficial microbes already present at the expense of pathogenic ones.

In one study in Hawaii, researchers found that introduction of just a small amount of microbe-rich topsoil from a healthy field reduced the incidence of root rot in papaya plants being grown on a primitive, volcanic soil. The volcanic soil had very little organic matter and relatively low soil biodiversity. Papaya seedlings planted into depressions filled with microbe-rich topsoil from the other site had significantly less root rot and higher yields.

THE OBSERVATION THAT SOILS can be made disease-suppressive by adding a relatively small volume of compost or soil from another site has led to investigations of the biological factors involved. Particular strains of bacteria in the genera *Bacillus, Streptomyces,* and *Pseudomonas* have been identified as the beneficials in many cases. In the late 1970s, Milton Schroth of the University of California at Berkeley was one of the first to show that inoculation of the roots of young potato, sugar beet, and radish seedlings with specific bacterial species in the genus *Pseudomonas* could significantly protect the plants from soilborne disease and increase final yields by 30 to 100 percent or more. The effect of these beneficial bacteria is usually long-lasting because their numbers continue to increase and they colonize new plant roots as they develop.

The primary benefit of the symbiotic mycorrhizal root fungi is enhancing a plant's access to water and nutrients (see Chapter 5), but sometimes they also protect roots from invasion by harmful microbes. This protection may be just the inadvertent result of outcompeting pathogens for sites on the plant roots.

A much more active form of protection of plant roots can be provided by another common type of root fungus, the nonmycorrhizal *Trichoderma harzianum*. The threadlike hyphae of *Trichoderma* actually target various disease-causing soil microbes and parasitize them. They first coil their hyphae around the victim, then penetrate the cell walls in order to suck out the vital fluids (figure 7.2).

The disease protection provided by *Trichoderma* is sometimes due to other mechanisms, such as the release of antibiotic-like substances into the surrounding soil or into the plant, or a general stimulation of plant growth. Once *Trichoderma* establishes itself within the root zone of a young plant, it continues to colonize on or near new roots as the plant grows, often providing lifelong protection from pathogens. Numerous greenhouse and field tests with *Trichoderma* have demonstrated its effectiveness at protecting plants from several soilborne pathogens and its positive effect on crop yields. It is primarily preventative, however, and not very effective against diseases that have already become well established on a plant.

Dr. Gary Harman of Cornell University isolated strains of *Trichoderma* from soils that were particularly good at suppressing soilborne pathogens. Utilizing a modern genetic engineering technique known as protoplast fusion, he created a superior new strain, T-22. Retail sales of T-22 products for commercial agriculture totaled around $3 million in 1999, and sales are expected to grow substantially over the next several years.

There are beneficial bacteria as well as fungi being used commercially as biocontrol agents of plant disease. These microbial pesti-

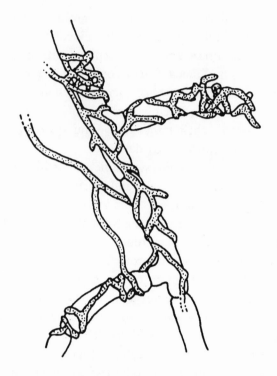

FIGURE 7.2 *The beneficial Trichoderma fungus coiling its hyphae around the wider hyphae of the pathogenic fungus Rhizoctonia solani, which causes damping-off disease. From J. W. Deacon, Microbial Control of Plant Pests and Disease (Washington, D.C., 1983).*

cides, as they are sometimes called, are generally considered much safer than synthetic pesticides because they target specific pathogens, their activity is confined to the root zone, the chemicals they release are organic, and the quantities released are very small compared to the quantities of toxins applied in traditional pest control. Biological controls therefore are promoted as particularly well suited for use in urban areas, homes, golf courses, parks, and other areas where environmental contamination by pesticides is of concern.

Microbial pesticides still must be registered for use by the Environmental Protection Agency in the United States, a process that can take three years or more. Some unscrupulous manufacturers get around this requirement by labeling their product a "growth promoter" rather than a "pesticide." This practice is of concern to legitimate companies and scientists working with biocontrols because the premature release of one microbial product that creates an environmental hazard or a human health risk would undoubtedly have a long-term negative impact on future development of genuinely safe and effective biocontrol products.

Moreover, some people are not reassured by EPA approval of a product, and many fears have been raised about the release of genetically modified organisms. Some of these fears are based on misinformation, but there are legitimate concerns, and we could always be blindsided by some unforeseen ecological disaster. The hard fact is that there are no guarantees. Science offers no certainties, only probabilities that always fall short of 100 percent. What are the long-term consequences of status quo synthetic pesticide use compared to expanded use of soil microbes? A decision must be made in the midst of uncertainty.

And then there are issues of environmental ethics. Is one approach to pest control ethically superior to another? Or is the moral high ground to ban all pesticides and allow food shortages to control the rate of human population growth in the future? To become paralyzed and do nothing because of scientific uncertainty and moral conflict is in itself a decision with consequences. Let's hope that our society does not shy away from these issues but actively debates them. One approach will be to develop, within the context of our best scientific information, regulations for cautious experimentation with genetically modified organisms.

ALTHOUGH THE USE OF SOIL MICROBES to fight plant disease is a relatively new option, we have been using soil microbes for several decades with great success to fight human diseases. Few people realize that most of the natural antibiotics in use today are produced by soil bacteria. Many of the same species of bacteria being raised in the sophisticated laboratories of pharmaceutical companies for commercial-scale production of antibiotics can also be found in your own backyard. And many of the man-made, or synthetic, antibiotics were not really invented by us from scratch but are modified versions of natural products of soil bacteria.

FIGURE 7.3   Dr. Selman Waksman, the soil biologist who discovered streptomycin and won the Nobel Prize in Physiology and Medicine in 1952. Courtesy of Rutgers University Archives collection.

One of the key scientists who helped to usher in the antibiotic era, which revolutionized medicine in the mid-twentieth century, was in fact a soil biologist, Dr. Selman Waksman, of the Agriculture Experiment Station at Rutgers University. Waksman discovered streptomycin and several other "miracle drugs" in his soil samples and was the first to use the term "antibiotic" in 1941 to describe these medical wonders.

Most of us are more familiar with the Scottish physician Sir Alexander Fleming, who discovered penicillin in 1928. Fleming's discovery was serendipitous, the outcome of accidental contamination of a petri dish by spores of the fungus *Penicillium notatum*. Fleming published several papers on his observations between 1929 and 1931, but little came of this work until, in 1939, a method was devised to produce enough penicillin to test its medicinal properties. Two Oxford scientists, Howard Florey, a plant pathologist, and Ernst Chain, a chemist, rediscovered Fleming's early work and devised the first clinical trials of penicillin on laboratory mice they had infected with a pathogenic species of *Streptococcus*. Nearly all of the mice treated with penicillin survived, while the others died, and the rest is history. (That history might have been very different had they chosen guinea pigs as their test animal, since penicillin, for some still unknown reason, is toxic to guinea pigs.) Fleming, Florey, and Chain received the Nobel Prize in Medicine for their collaboration in 1945.

During the mid-1930s, while progress with penicillin was still in limbo, Selman Waksman was beginning his search for antibiotics in the soil. He had always been intrigued by the disease-suppressive capacity of soils. This interest had its roots in his agricultural background. As a young boy growing up in a rural area of the Ukraine, and later as a young man working on a farm in the United States while attending college, he had been fascinated by how soils seem to self-purify, recovering year after year from the incorporation of dead and diseased plants and animals.

As a pioneer soil ecologist, Waksman discovered that microbes produce both growth-promoting substances, such as vitamin B-12, and growth-inhibiting compounds (which he called antibiotics) that affect their interaction with each other. Waksman noted that some of the antibiotics he isolated from the soil were effective against close relatives of species pathogenic to humans. The first breakthrough came in 1932 when one of his students, René Dubos, discovered a compound produced by a soil bacteria that inhibits a bacteria responsible for pneumonia.

During 1940 and 1941, great progress was made in the development of penicillin, but it was effective only against the so-called gram-positive bacteria. It proved of no help in stopping one of the major killers of that period—the "white plague," or tuberculosis. In the meantime, however, Waksman's group had begun a collaboration with the Merck pharmaceutical company and was coming up with a slow but steady stream of promising new antibiotics from the soil. Many of these were from very common soil bacteria in the group known as actinomycetes, which produce a wide array of compounds, not all of which are antibiotics. The earthy odor of moist forest soils, for example, is caused by a volatile substance, geosmin, produced by actinomycetes. In 1940 Waksman released his first antibiotic isolated from these bacteria, aptly naming it actinomycin. This was followed by the discovery of clavicin, fumigacin, and streptothricin in 1942. Although each of these antibiotics held promise, each also had flaws, either in their effectiveness in fighting disease or in their toxicity to animals.

Finally, in 1943, Waksman isolated an antibiotic produced by the actinomycete *Streptomyces griseus* that was nearly perfect. In laboratory tests, it was highly effective against many of the pathogens that penicillin could not inhibit, and trials with human patients at the Mayo Clinic soon proved that it was at least partially effective against tuberculosis. Selman Waksman named this new miracle drug streptomycin, and it soon was front-page news and the subject of a *Time* magazine article on January 29, 1945. By 1946 streptomycin was being commercially manufactured and saving many human lives from forms of tuberculosis that were previously incurable. Within a few years, streptomycin was being used with great success to combat other diseases and infections that had not been responsive to penicillin therapy, including tularemia, plague, meningitis, brucellosis, and many forms of colon and urinary tract infections. Waksman had discovered the antibiotic the world needed, and armed with both penicillin and streptomycin, doctors had an unprecedented capacity to cure deadly diseases.

Selman Waksman went on to develop other antibiotics, but none surpassed streptomycin in terms of beneficial impact. Unlike the accidental discovery of penicillin, Waksman's discovery of miracle drugs from the soil was a deliberate effort and the result of many years of meticulous and focused labor. In his autobiography, Waksman shares credit with his students and attributes much of his success to the academic freedom afforded him by the institution where he spent his career, Rutgers University. In retrospect, he was as surprised as anyone that a soil biologist working at a small agricultural lab would end up playing such a major role in curing human disease.

Like all scientists who have pioneered new fields of study, Waksman had many detractors along the way. His applications for federal funding for his work were often rejected on the grounds that his proposals were too basic and theoretical. In his autobiography, he recounts being told by one prominent scientist in Washington, D.C., that there was no branch of science that had yielded so little information of practical value as had soil microbiology. One can only wonder what that scientist thought when, in 1952, it was announced in Stockholm that Selman Waksman would be that year's recipient of the Nobel Prize in Physiology and Medicine.

WE MUST LEARN TO RECONCILE the dual nature of soils. On the one hand, among the tens of thousands of species that live in the soil, a few are indeed capable of turning on us or our food crops and becoming pathogenic under certain environmental conditions. On the other hand, the vast majority of soil microbes are either harmless or directly beneficial to us. Many of our most potent antibiotics for fighting human disease are derived from the soil, and farmers are learning how to put soil creatures to work in combating food crop diseases. The fact is that we are part of a complex food web in which nearly every creature has its natural enemies to contend with and simultaneously can become the enemy of certain other species. We can minimize negative encounters with our potential enemies that live in the underground by proper hygienic care of open wounds, immunization against tetanus, and careful management of the soils in which we grow our crops.

Protection of soil biodiversity should be a high priority, for who knows which of the many thousands of genetic types of microbes living in the underground might provide us with the next miracle drug or become a new reliable biological agent for the control of plant disease? As we shall see in the chapters to follow, minimizing our impact on creatures of the underground is likely to be a bigger challenge than minimizing their impact on us.

Part Three

THE HUMAN FACTOR

8

ENDANGERED DIGGERS
OF THE DEEP

*Wouldn't it be great if we could declare a ceasefire
in our century-old war against prairie dogs? I'm
convinced there's enough room for both people
and prairie dogs in the American West.*

—DAVID WILCOVE, SENIOR ECOLOGIST,
ENVIRONMENTAL DEFENSE (2000)

AT THE BEGINNING OF THE NINETEENTH CENTURY, Meriwether
Lewis and William Clark set out on their famous exploration of the
American West. The "Corps of Discovery" that they led was commis-
sioned in 1803 by Thomas Jefferson, then president of the United
States. Their mission was to map the primary waterways from St.
Louis, Missouri, to the Pacific Northwest and, just as important, to
describe and catalog the natural wonders they encountered.

Lewis and Clark filled their journals with detailed descriptions of
thousands of new plant and animal species. What they could not
have foreseen was that western settlement, which their expedition
was designed to encourage, would bring a number of these unique
native species to the brink of extinction. Three of the affected
species—prairie dogs, black-footed ferrets, and burrowing owls—are
the focus of this chapter. The demise of prairie dogs in particular is
of concern to ecologists because prairie dogs are a "keystone" species
and, as such, act as the glue that holds the vast grassland ecosystems
of the American West together. The decline in their numbers has had

a tremendous impact on many plant, animal, and microbial species, both above- and below-ground.

LEWIS AND CLARK HAD LITTLE reason to suspect that the life beneath their feet was of much significance. Their expedition took place, after all, almost three-quarters of a century before the publication of Darwin's book on earthworms, which represented a turning point in our appreciation of subterranean life. Without microscopes, the explorers' observations of life in the underground were limited, but one charismatic creature could not be missed. This was the burrowing mammal that they referred to as the "barking squirrel" and we know as the prairie dog (*Cynomys* species). Lewis and Clark were not the first white men to see prairie dogs (French explorers before them had talked of the *petit chien,* or "little dog"), but they were the first to examine the animals carefully, and the first to enter a formal description into the annals of science.

Their first encounter was on September 7, 1804. After a cold morning of rowing several miles up the Missouri River to a point within present-day Boyd County, Nebraska, they put ashore at the base of a small hill. While some of the men rested and rearranged things on the boats, Lewis and Clark set out on a leisurely stroll together, something they did not often have the opportunity to do. They walked to the top of the hill to have a broader view of the terrain, then meandered down the other side. They had not quite reached the base of the hill when they noticed something very peculiar about the landscape. Wary travelers in unknown territory, they stopped, turned slowly to scan the area, and found that they were surrounded by hundreds of small grass-covered mounds, encompassing about four acres. The grass was thick and evenly clipped, like a well-manicured bowling green. This was the work of either humans or some large, highly organized community of unfamiliar animals. As able hunters, their sixth sense told them that they were not alone, and soon they noticed signs of movement everywhere around them.

The tiny brown eyes of hundreds of sandy-colored squirrel-like rodents were on them. The animals were clearly as startled and mystified as were Lewis and Clark. Some of the animals were peeping up from burrow entrances, while others boldly stood atop the mounds, up on their hind legs, using the vantage point to get a better look at the intruders. In semiferocious stance, a few of the animals challenged the explorers with their "teeth-chatter" display. Others let out a high-pitched bark, piercing the silence, while throwing their heads and forefeet back in the "jump yip" territorial display. Within the blink of an eye, word of danger had spread like a wave throughout the entire "village" (as Clark would later describe it in his journal), and most of the strange creatures had vanished. Silence again. The captains rubbed their eyes. One of them took a long stick and tried, unsuccessfully, to prod one of the residents out of its burrow. Frustrated, the captains hiked back to the boats and ordered a couple of men to grab shovels and join them. A decision had been made. They would not be moving on until they had captured some of the unusual subterranean creatures for closer examination.

What they thought might take just an hour or so ended up occupying the entire day and required the assistance of almost the entire crew of the Corps of Discovery. Their poking and prodding with sticks of all lengths proved futile, as did attempts to dig the animals out. The burrows were too deep and intricate. Finally, in desperation, they used all the barrels and other containers they had available to haul water to the site, which they poured down one of the holes. This too seemed to fail until, just as the sun was setting, a drenched and exhausted resident emerged from its underground home. The hapless creature was immediately sacrificed for purposes of examination and classification.

We can now say that this was an omen of things to come for the prairie dog, whose future from this point on would be in the hands of the white man and the "manifest destiny" of westward expansion. Lewis and Clark could not have envisioned, of course, the mass prairie dog eradication programs that would be initiated later in the nineteenth century. For them at the time, collecting animal specimens was an obvious necessity for fulfilling their scientific mission. Lewis's detailed description would be the first published account of this subterranean mammal.

In the days that followed, the Corps of Discovery came across other prairie dog "towns" much more immense than the first. In Lewis's journal entry for September 17, 1804, when they had just crossed into what is now South Dakota, he describes a vast region several miles wide and "intirely occupied by the burrows of the barking squiril heretofore described; this anamal appears here in infinite numbers and the shortness and virdure of grass gave the plain the appearance throughout its whole extent of a beatifull bowling-green in fine order."

With each new dog town they came across, Lewis and Clark became more fascinated with these highly social creatures of the underground, and they would spend considerable time studying them. In another entry, after a detailed description of the morphology and dimensions of individual prairie dog specimens, Lewis writes:

> These animals form in large companies, occupying with their burrows sometimes two hundred acres. . . .[Their] mounds, sometimes about two feet in height and four in diameter, are occupied as watch towers by the inhabitants. . . . When anyone approaches, they make a shrill whistling sound, somewhat resembling tweet, tweet, tweet, the signal for their party to take the alarm, and to retire into their intrenchments.

Recent research by Dr. Constantine N. Slobodchikoff of Northern Arizona University suggests that prairie dogs not only have unique alarm calls for different predators but include descriptive "words," if you will, in their vocalizations. In one intriguing study, Slobodchikoff made recordings of alarm calls as humans of varying sizes and shapes walked through a prairie dog colony on three separate days. He also had the walkers alter their clothing. They wore white lab coats on some of their walks and brightly colored T-shirts on others. To the amazement of the researchers, detailed analysis of the sound wave frequencies of the recordings indicated that the prairie dogs not only had a specific alarm call for humans, but encoded information into their vocalizations about the color of the clothes and general shape of the humans eliciting the calls. At the sighting of Lewis and Clark, the prairie dogs may have added to their lexicon the equivalent of "white man"—a phrase that was destined to become associated with death and destruction for their species.

Throughout the nineteenth century, explorers, naturalists, and U.S. Army generals filled their journals with accounts of prairie dog communities of incredible size and attempted to describe the complex social behavior they observed. One of these was General Zebulon Montgomery Pike, who was exploring parts of the Arkansas River basin at about the same time Lewis and Clark were traveling the upper Missouri. Another was the naturalist painter John James Audubon, who updated the scientific description of various prairie dog species in the 1850s. General George A. Custer took time out to write about prairie dogs in his field journal on his way to the Little Big Horn River of Montana in 1876.

Some of the dog towns encountered by settlers heading west were so immense that it took the travelers several days on horseback to traverse them completely. One in Wyoming was 100 miles (160 kilometers) long. The real whopper, the one for the *Guinness Book of World Records,* was a town in Texas that occupied 25,000 square miles (about the size of West Virginia) and may have contained 400 million individuals. The estimated total prairie dog population in North America in the mid-nineteenth century was five billion. Their communities were commonly found over a wide range, from central Canada down through Texas to northern Mexico, and west to the Rocky Mountains.

THE JOURNAL ENTRIES OF MERIWETHER LEWIS reflect an ecologist's sense of the importance of food webs and interactions between plant and animal species. In his September 17 entry, for example, he noted that huge herds of deer, elk, pronghorn (*Antilocapra americana,* which resemble antelope), and buffalo roamed and were nourished by the lush grassland being groomed by the prairie dogs. He deliberately searched the area for evidence of likely predators of the prairie dog. He must have been wondering what constrained the continued expansion of these vast "barking squiril" colonies. He wrote that "a great number of wolves of the small kind, halks, and some polecats were to be seen. I presume that those anamals feed on this squiril."

Today we know that, as a keystone species within their natural habitat, prairie dogs are similar to earthworms, nitrogen-fixing bacteria, and some of the other species of the underground discussed earlier. Remove any of these keystone species from an ecosystem, and things quickly fall apart, to the detriment of many other plant and animal species with whom they interact. Prairie dogs were arguably the most essential ingredient in the highly diverse and productive grasslands that once occupied nearly one-fifth of the North American continent.

Prairie dogs are sophisticated landscape managers, and many other species benefit from their work. They keep tall grasses and developing shrubs clipped back to a uniform height so that they can easily spot intruders. This constant mowing of the vegetation promotes plant diversity and leaves behind an abundance of succulent, shorter grass species, including the protein-rich clovers and other legumes preferred by grazing ungulates (hoofed animals), such as buffalo and domestic cattle. In the winter, the diet of prairie dogs in many areas is predominantly prickly pear cactus (*Opuntia plycantha*), a plant that cattle do not eat and that often takes over in cattle-grazed areas where prairie dogs are not present. The woody mesquite tree is another plant species that can become dominant in grasslands that have no prairie dogs. These small trees shade out grasses and other forage species and make cattle roundups more difficult for ranchers.

Prairie dogs begin their activity at sunrise and spend several hours feeding on vegetation, manicuring the landscape, basking in the sun atop their mounds, and grooming each other (presumably) to remove ticks and fleas. The majority move underground for the heat of the day, reappearing for a while near sunset, and then retiring for the evening.

Almost half of the animals' waking hours are devoted to scanning for predators, such as badgers, coyotes, foxes, ferrets, bobcats, rattlesnakes, eagles, falcons, and hawks. Mounds are used as lookout posts. It is no easy trick for a predator to sneak into a colony without being seen by a prairie dog who will immediately send out the alarm. Nevertheless, stealthy predators do succeed now and then, and thus prairie dogs play their role in the food chain. Some species of prairie dog are active in winter, providing an essential source of nourishment for predators during a period when most other prey are hibernating. Prairie dog colonies support much larger populations of predators than are found in the surrounding grasslands.

The social structure that holds the large prairie dog colonies together is nearly as complex as that observed for wolves, dolphins, and primates. One species, the black-tailed prairie dog (*Cynomys ludovicianus*), has been particularly well studied, most recently by Dr. John Hoogland from the University of Maryland. The work of Hoogland and others has shown that large dog towns or colonies are subdivided into wards, which are further divided up into small social units called coteries. The borders of wards are usually defined by physical features of the terrain, such as streams, rock outcroppings, or vegetation. Within a ward, there may be anywhere from a few coteries to dozens of them. A coterie is a family of closely related

FIGURE 8.1   *The characteristic "kissing" greeting between two prairie dogs of the same family or coterie. Courtesy of Tom and Pat Leeson.*

individuals, typically with three to four breeding females, one breeding dominant male called the "harem master," and several nonbreeding yearlings and juveniles.

Each coterie occupies and defends its territory and burrow network, which may occupy a little less than an acre and have fifty or so burrow entrances. Members of the same coterie acknowledge and greet each other with a characteristic "kissing" behavior (figure 8.1). But when a male from one coterie tries to enter another's territory, the battles can become vicious. They usually avoid serious injury in these standoffs, however, by stand-offs that involve a lot of staring, teeth-chattering, and bluff-charges.

Inbreeding is taboo for prairie dogs. The dominant male moves on to search for a new harem and coterie when his daughters reach sexual maturity at about two years of age, making sexual relations between close relatives a rare event. With the moving about of the males, the occupants and genetic background of coteries change with time, but the physical boundaries usually do not. Males typically live about five years and females about eight.

There is a dark side to prairie dog coloniality and social life. It is sometimes the case that when two or more females are raising young simultaneously within a coterie, one female may enter the nesting area of the other and kill some or all of her babies. By killing another female's young, a mother eliminates competitors with her own offspring. Although females defend their young against such attacks, once the deed is done they do not hold a grudge for long. Paradoxically, communal nursing and care of young is also very common, and in fact frequently observed in a coterie where infanticide has recently occurred.

Communal living also means that diseases can be easily spread. Prairie dogs are particularly susceptible to bacterial plague (*Yersinia pestis*), which is usually carried by fleas or ticks. This disease has not been a serious threat to humans since the Dark Ages, when it was known as bubonic plague, or the "Black Death." Today humans at risk can be immunized, and there are effective antibiotics for the handful of individuals who become infected each year. Prairie dogs, however, are still living in the Dark Ages when it comes to plague, and entire colonies can be devastated within a short period of time.

It was our species that brought bubonic plague to North America a hundred years ago or so, carried by fleas on animals unloaded from European (or perhaps Asian) ships. The immigrants to the New World also brought with them smallpox and other human diseases that killed hundreds of thousands of Native American Indians, who, like the prairie dogs, lacked resistance. But the impact of these diseases on the native humans and animals of the Americas was eventually overshadowed by the full-fledged war waged against them both by the European settlers.

UNFORTUNATELY, NONE OF US WILL ever have the opportunity to see a dog town stretching far into the horizon. A firsthand glimpse of such a sight, the kind of experience that had a seasoned explorer like Meriwether Lewis reaching for his journal, is a thing of the past. Prairie dog communities, like the Native American communities that had co-existed with them for centuries, fell victim to ambitious westward expansion by American settlers. Small dog towns can still be found here and there throughout the Great Plains and Rocky

Mountains, but the population has been greatly diminished and highly fragmented. Prairie dogs today are only safe within federally protected lands. They occupy less than 2 percent of the land area they once did within North America, and during the past century the total population has been reduced by more than 90 percent.

The farmers and ranchers who took up residence in the Great Plains viewed the prairie dog, whose habitat they had invaded, as public enemy number one. With their greatly exaggerated notions of how much vegetation the prairie dogs consumed, the ranchers became convinced that the animals had to be completely eradicated if cattle were to be raised on the same land. Poisoning campaigns were initiated in the pioneer spirit of "conquering" nature.

The ecology of the area—the fact that for centuries huge herds of buffalo, elk, pronghorn, and deer had thrived on the grasslands groomed by prairie dogs—was ignored. Prairie dog colonies, when held in check by predators in a well-balanced ecosystem, can pro-mote a particularly succulent and nutritious forage preferred by domestic cattle as well as wild species. Such benefits of prairie dog presence were never taken into consideration.

By the 1880s, ranchers had settled on strychnine-laced oats as the weapon of choice in their battle with the prairie dog. The prairie dogs were so numerous, however, that it proved difficult and expensive to eradicate them all. By the turn of the century, some ranchers were beginning to question the economics of the war on prairie dogs, but in 1902 C. H. Merriam, director of the U.S. Biological Survey (pre-cursor to the U.S. Fish and Wildlife Service), gave official support to the poisoning campaigns. Basing his claims on hearsay rather than scientific evidence, he publicly announced that prairie dog colonies were reducing the productivity of rangelands by 50 to 75 percent. Recent research indicates that this was a tenfold exaggeration fanning the flames of rancher animosity toward the prairie dog.

The Great Plains region was gripped by a warlike mentality toward prairie dogs throughout the early 1900s. Proposals for more meas-ured, rational management strategies were ignored. It is easy to look back and be critical, but if you have ever been frustrated in your efforts to control a household or garden "pest," you can probably appreciate how quickly this struggle escalated into an obsession. The ranchers of the Great Plains thought their very livelihood was at stake. Unfortunately, their battle was played out on a much grander scale than a home garden. It engulfed an entire grassland ecosystem that had once occupied nearly one-fifth of North America.

During the early 1900s, natural prairie dog habitat was being systematically destroyed, and many landowners were spending so much on poisons in the process that they were approaching economic ruin. In retrospect, the ranchers were less to blame for this folly than the public officials who failed in their responsibility to step back and evaluate the situation with cooler heads, and with an eye to long-term preservation of our natural resources.

By 1915 the federal government (that is, U.S. taxpayers) had begun footing the bill for 75 percent of poisoning expenses, thus making the eradication campaign "cost-effective" for the ranchers. Federally funded research on prairie dogs shifted away from ecological studies, which might have led to a more rational policy, and instead began to focus almost exclusively on improving the efficacy of poisoning methods. By 1920 the U.S. Biological Survey was in the business of killing millions of prairie dogs each year. In that year alone, 1,610 tons of strychnine-coated grain were applied on 18 million acres of ranch and farm land, and on 4.5 million acres of public land. By 1925 an entire new department, the Division of Predatory Animal and Rodent Control, had been formed to carry on with this enterprise.

This concerted effort brought some "success." The total area covered by prairie dog colonies was reduced from somewhere between 100 and 250 million acres (40 to 100 million hectares) in 1870 to less than 1.5 million acres by 1960. The colonies that have survived the combined effects of poisoning and disease are minuscule in comparison with what the early settlers witnessed, and their geographical isolation threatens species survival in some regions. Were it not for the refuges within the U.S. National Park system and tolerance of prairie dogs in land areas unsuitable for ranching, prairie dogs might now be extinct.

Recent cost-benefit analyses have shown that poisoning programs operate at a net financial loss. They continue, however, even on federally leased land, because of government subsidy and lack of trust by landowners in other alternatives. As a result, America's prairie grasslands have become an ecosystem out of balance. The solution may lie in avoiding overgrazing by livestock while allowing expansion of some prairie dog colonies to a size that would once again attract and support a population of their natural predators. Subsidizing ranchers

FIGURE 8.2 *A photograph taken in the early 1900s showing a pile of dead prairie dogs left after the poisoning of a small colony. Courtesy of Predator Conservation Alliance.*

for the economic risks associated with such a shift in land management policy would probably be cheaper than a never-ending campaign aimed at complete eradication of a keystone species.

There are some signs that attitudes are changing. In July 1998, the National Wildlife Federation and the Predator Conservation Alliance requested that the black-tailed prairie dog be listed as "threatened" under the Endangered Species Act. The preliminary U.S. Fish and Wildlife Service decision on this matter, announced in February 2000, was that the request was "warranted" but could not be granted immediately without additional information. In the interim, the agency agreed to put protective measures in place on public lands and to review the species' status annually.

AS DEVASTATING AS THE POISONING of the Great Plains has been for the prairie dog, the impact has been even more severe for several other burrowing species that depend on the prairie dog for their survival. Undoubtedly the hardest hit has been the black-footed fer-ret (*Mustela nigripes*), a highly specialized predator of the prairie dog (figure 8.3). Several times since 1950, scientists have been ready to throw in the towel and declare this species completely extinct. Each time there has been a reprieve, a new enclave discovered some-where, or some success in captive breeding. It has been estimated that in the mid-nineteenth century, there were close to one million black-footed ferrets co-existing with the prairie dog; by the end of the twentieth century, there were probably no more than a few dozen. Today it is quite possibly the rarest of all mammals native to North America.

The black-footed ferret is among those predators whose primary prey is as big or bigger than it is. But everything about this ferret is specifically designed for the job it needs to do. Beneath a sweet and endearing exterior is a hunter with incredibly strong jaws, a long, supple body for maneuvering in the subterranean, and a keen sense of smell. In the winter, it can easily find burrow entrances buried under a foot or more of snow. Its spine is so flexible that within a tight tunnel it can simply roll and walk back over its own hindquar-ters as though made of rubber.

The method of attack and the "killer bite" that black-footed ferrets have perfected are enough to make one shudder. In the dead of night, the ferret creeps through the coterie burrows until it finds a lone prairie dog sleeping. Rather than pounce immediately, it first care-fully positions itself just a few inches from the victim. When every-thing seems just right, it reaches out one forepaw and gives the prairie dog a light touch or two on the shoulder—*tap, tap.* As the drowsy prairie dog wakes from its slumber and lifts its head to see what is going on, the ferret snaps violently, plunging its long canines into the dog's neck.

A 1,000-acre (400-hectare) prairie dog colony may hold several thousand prairie dogs, but this will only be able to support five to ten breeding pairs of black-footed ferrets. To prevent inbreeding, the fer-rets need to be able to reach other ferret populations living in other prairie dog colonies. Unfortunately, most prairie dog colonies today are too small and fragmented to meet the criteria for ferret long-term survival. As good as the ferrets are at killing prairie dogs, they are not very competitive as predators of alternative prey—thus their demise.

FIGURE 8.3 *A black-footed ferret, one of the rarest mammals of North America. Courtesy of Dean Biggins, U.S. Geological Survey, Biological Research Division.*

Between 1946 and 1953, there were only seventy black-footed ferrets seen in the wild, despite some deliberate efforts to seek them out. By the mid-1960s, many biologists were prepared to declare the animal extinct. When the U.S. Endangered Species Act was passed in 1966, the black-footed ferret was among the first to be put on the list. During the 1970s, there were very few sightings, but the U.S. Fish and Wildlife Service was attempting to preserve a few animals by raising them in captivity. This effort created dueling governmental programs—one in the business of destroying the ferret food supply (prairie dogs), and the other involved in a desperate attempt to save the ferret from extinction. Not surprisingly, this schizophrenic approach met with little success.

Most thought the story was over when the very last captive ferret died of cancer in 1979. Then, one September night in 1981, in Meeteetse, Wyoming, an old ranch dog named Shep killed a ferret that he found eating at his food bowl. Shep's owners, John and Lucille Hogg, were impressed by the unusual pelt of the dead animal and took it to a local taxidermist to have it preserved. The taxidermist recognized it immediately as the rare black-footed ferret and called the authorities.

Within days, a few of the most dedicated scientists from government agencies and universities arrived in the area to search for the relatives of the ferret that had been killed. Despite a lack of funding, they spent night after night methodically sweeping the nearby prairie dog habitat with flashlights "spotlighting"—looking for the recognizable reflection of black-footed ferret eyes. Before winter set in, they had verified that there was a viable and significant population in the area. Americans had not yet completely eliminated this wonderful piece of their natural heritage.

Unfortunately, many members of the small population of ferrets discovered near Meeteetse fell victim to either canine distemper or plague in 1985. Meetings were hastily called together, but there was no clear agreement on what to do. A great deal of precious time was wasted by indecision and debate among the university and government agency personnel. Should they try to capture every ferret they could? Should they leave them in the field and drench the area in pesticides to kill the flea disease vectors? Finally, when the count had dropped to ten ferrets, six of the animals were brought into captivity and four left in the field. In the end, the scientists had four healthy females and two males from which to rebuild. It had been a terrible summer, but at least the ferret was not extinct.

During the late 1980s and early 1990s, dozens of public and private groups became involved in the effort to preserve this rare species. Although this was in general a positive development, there was no one group or agency with clear authority to make decisions, and the various parties involved began to compete for the right to raise the captive animals. Parochialism erupted. Some Wyoming officials were reluctant to let any black-footed ferrets leave their state, even though there were better facilities to house the animals in Virginia.

With time, many of these issues were resolved, and a better level of cooperation developed among the many well-meaning groups. As of the spring of 2000, at least thirty-five private and government agencies are working together to raise and reintroduce captive ferrets into the wild. These groups have put to use what we have recently learned about captive breeding, disease management, behavioral conditioning of the animals for reintroduction, and prairie dog politics. Half the captive population currently resides at the National Black-Footed Ferret Conservation Center, which has been established near Meeteetse, Wyoming. The rest have been placed at the National Zoo Conservation Research Center in Front Royal, Virginia, or at one of several other zoos and conservation centers around the United States. The goal is to maintain a healthy captive population of about 240 individuals, with minimal inbreeding, and to reintroduce ferrets into the wild in several locations each year.

Black-footed ferrets have recently been reintroduced in Wyoming, South Dakota, Montana, Utah, and Arizona. The long-term success of these animals is still uncertain. Captivity obviously leads to an erosion of the skills needed to survive in the wild. Most biologists are quite optimistic that special methods, such as exposing the ferrets under controlled conditions to prairie dog colonies and potential predators like the badger, can prepare them adequately. A few scientists remain skeptical that these behavioral conditioning methods will work in the long run. Only time will tell.

The loss of survival skills may be the least of the black-footed ferret's problems. A minimum viable population of ferrets requires a prairie dog colony with at least several thousand prairie dogs. With continued poisoning on both private and federal lands, colonies of this size are becoming increasingly rare.

FOR THE FIRST SETTLERS in present-day New Mexico, the Zuni Indians were a great source of information. The settlers heard many tales of local wildlife, tales that interwove myth and reality. One of these tales was of a creature of the underground, a unique small bird that lived together with the prairie dog in its burrows. The Zunis called this bird the "priest of the prairie dog."

Lewis and Clark had also heard stories about various animals living harmoniously with the prairie dog, but the bird the Zunis spoke of, known now as the burrowing owl (*Athene cunicularia*), never made it into their journals. This is surprising because this unique and diminutive member of the owl family, small enough to fit in the palm of your hand, is common wherever there are prairie dogs (figure 8.4). Perhaps the records were lost, or perhaps Lewis and Clark really did somehow miss out on one of the most intriguing animal interactions on the prairie.

The burrowing owl's common name is misleading because for the most part this species is incapable of digging its own burrows. The owls survive as burrow "borrowers": They move into abandoned nests constructed by prairie dogs or other animals of the underground. Prairie dog habitat is prime real estate from the owl's perspective because, in addition to the well-ventilated subterranean quarters, prairie dogs provide superior aboveground lawn service and mounds that make excellent perches. Standing atop the mounds near the burrow entrances during the day and twilight hours, the owls can watch for predators and scout for insects and other food items. It is thought that the prairie dogs and owls mutually benefit from each other's presence by taking turns as "lookout" for predators. The owls are not nocturnal like other owls, but they tend to wake up earlier and go to sleep much later than the prairie dog.

The burrowing owl has a broader geographic range than the prairie dog or black-footed ferret, encompassing much of western North America as well as parts of Central and South America, Florida, and the Caribbean. The Chilean priest who first described the birds in 1782, Father Molina, came up with the species name *cunicularia,* from the Latin for "rabbit-like," suggesting that the owls he happened to observe were sharing subterranean space with rabbits. In the absence of prairie dogs, the owls can sometimes make do with the burrows of other animals. Ground squirrel nests are a second favorite to those of prairie dogs. The owls also use burrows dug by rabbits, marmots, kangaroo rats, and armadillos. The owls have even been known to boldly move into burrows abandoned by their most feared predator—the badger. The owls cover the walls and entrances of the badger's burrow with the feces of cattle or other animals, a tactic that seems to effectively keep the owner away.

The only burrowing owls capable of digging their own nests are a subspecies in Florida. Why this is so remains a mystery. It may be that digging in the soft sandy soils of Florida is easier, or that there are fewer predators (so nests do not need to be very deep). Or perhaps

this owl has managed to pass on burrowing know-how from generation to generation more effectively than its relatives in other parts of
the Americas.

Burrowing owls are typically monogamous within any one breeding season. There is none of that lustful harem business observed
with prairie dogs. Newly paired owls can be seen out shopping for
their first underground home in the late winter or early spring, just
before the mating season. Once in a while they may aggressively evict
a prairie dog or ground squirrel to gain a piece of real estate they
want, but more often they choose abandoned burrows. What the
owls are looking for in a home is roomy and comfy subterranean
space, low vegetation near the entrance, a mound or perch so they
can see any predators coming, and nearby feeding grounds.

Favored prey are insects, mice or other small rodents (prairie dogs
and ground squirrels are much too big), and small reptiles or birds.
The amount of open grassland required by the owls varies widely,
from several acres to hundreds of acres in arid regions with low food
supply. Burrowing owls are semicolonial: They like being near a few
other owl families, but not too many, and not too close. They adapt
easily to human-altered landscapes; small colonies have been found
inhabiting golf courses, parks, airports, vacant lots, and freeway
cloverleafs.

Once a burrow has been selected, the male often adorns the
entrance with decorative items, such as feathers, beetle wings, and—
in urban areas—tin foil, shredded fast-food containers, or cigarette
butts. Once a couple moves in, they seldom change residences unless
the burrow is destroyed. This burrow fidelity is even seen in regions
where the owls migrate south for the winter. When a couple returns
to the northern breeding grounds after their winter hiatus, they go
right back into the same summer home they have always enjoyed,
provided it is still there. Mated pairs or parents and their young are
often seen standing, cheek to cheek, near the entrance to the nest
(figure 8.4).

During breeding season, the male owl performs a number of feats
to inspire the ardor of his mate. He demonstrates his aerodynamic
prowess in the "courtship display flight." This begins with a rapid
ascent of ninety feet (thirty meters), followed by a five- to ten-second
hover, then a daredevil nosedive halfway back to the ground before
making another rapid ascent. This is repeated several times, some-

FIGURE 8.4 A burrowing owl pair standing near the entrance to their burrow. Courtesy of Phyliss Greenberg.

times with rapid circular flights thrown in for good measure. If the female is impressed, she may reward the exhausted male with a little billing and preening in the head and facial area.

A good singing voice is another attribute that females look for in a potential mate. Burrowing owls do not "hoot" like other owls, but all males develop their own version of a "primary song." Male owls use this song for serenading as well as to announce their territory to other males. If the voice does not seem to be capturing the female's fancy, the male resorts to gifts, usually delicacies such as a fresh tasty toad or a plump juicy insect larvae.

Just before mating, both sexes display the "white and tall" stance to each other. They fluff up to reveal their white facial patches of feathers, while the male stands taller and more erect than the female, looking down at her with his best bedroom eyes. The sex act itself often occurs in the privacy of the nest but is also sometimes observed aboveground, especially at dusk. After copulation, the male sings a celebratory "coo-coo" song.

If all goes well, the female will soon be sitting on five to ten eggs, which is the largest clutch size of any raptor. The chicks hatch in just under a month. The male keeps mother and chicks supplied with food and fiercely protects the nest area from all intruders. If another owl or other animal gets too close, the male bobs up and down while declaring his turf with his primary song. If this has no effect, he fearlessly "presents" himself by spreading his wings. When all else fails, the male dives at the intruder, jabbing with his beak or claws. If the young in the burrow feel threatened, they may cry out with the "rattlesnake rasp," so named because it mimics the rattling sound of the deadly snake. This distress call may explain the (incorrect) legend that burrowing owls share their nests with the rattlesnake.

Despite the best efforts of the parents, on average usually only about one-third of fledglings survive to become reproducing adults. For those owls that make it to adulthood, the average life expectancy is about five years.

Burrowing owl habitat has been declining steadily ever since Europeans began settling in the Americas. The greatest loss, of course, has been the 98 percent reduction in prairie dog–engineered landscapes. Further loss of habitat and owl populations has accompanied agricultural expansion and urbanization in other parts of their range. The plowing and opening up of new farmlands destroys thousands of subterranean breeding grounds every year. Agricultural pesticides also cause losses, often associated with eggshell thinning and impaired development of the embryos. In addition, the owls suffer indirectly from the continued poisoning of prairie dogs, ground squirrels, or other animals they depend on for burrow construction. In urban areas, the problem is human disturbance and the literal paving over of nest sites. Collisions with vehicles and even deliberate shooting take a heavy toll on the burrowing owl in some areas.

The burrowing owl has completely disappeared in some parts of South America and is listed as "endangered" in Canada and in Minnesota and Iowa in the United States. In California and many other western states, almost none of the original prairie habitat remains, and burrowing owl numbers are much lower than they were a century ago.

The fate of burrowing owls in California, where owls have "special concern" status, may well be an indicator of things to come in other rapidly developing areas. Since California has no prairie dogs to speak of, the owls there usually occupy abandoned burrows of the California ground squirrel (*Spermophilus beecheyi*). The owl has adapted to living in human-altered landscapes, but it is a constant struggle for survival that gets more difficult with each passing year. It is estimated that 80 percent of the approximately nine thousand breeding pairs in California live within the two most active agricultural regions—the San Joaquin and Imperial Valleys. In those regions, the plow and pesticides combine to reduce and weaken the population. The famed Silicon Valley near San Francisco has historically been an important nesting area, but the owls are losing ground to new buildings and parking lots with each new computer or Internet company that sets up shop.

Dr. Lynne Trulio, a professor at San Jose State University, is an expert on both the California ground squirrel and the burrowing owl and has been at the forefront of owl preservation efforts in Silicon Valley for the past ten years. Despite her valiant efforts, however, and the efforts of other scientists and many concerned citizens, the owl population has declined by 50 percent during that period. In practice, the "special concern" status has provided insufficient protection for the owls. It does not prevent destruction of habitat; it only requires some effort be made at mitigating the impact of development on the owls. The mitigation efforts often are half-hearted and fail.

The struggle in Silicon Valley has had its ups and downs. Some public and private landowners have been very supportive of owl protection. For example, Trulio received a phone call from officials of Moffett Federal Airfield several years ago, asking for her help in managing the significant burrowing owl population that lives on Moffett's nine hundred acres of grassland. For Trulio, this was a real switch after working with nearby city officials who only wanted to eliminate the "owl problem" in order to pave the way for development. Since 1992, the U.S. Navy and NASA officials at Moffett have worked with Trulio to protect the owls and have funded some of her research.

In contrast, there have been many discouraging tragedies. One such incident occurred just a few blocks from Trulio's home. Plans were announced for development of a fourteen-acre lot occupied by three breeding owl pairs that she was studying. Many local citizens were concerned because they had come to know these owl families— some even considered them neighborhood pets. The local Audubon Society approached the city council, requesting that development on seven acres of this land, a section containing the owl nests, be restricted or at least delayed until a mitigation strategy could be arranged (as required by the bird's "special concern" status). More than one hundred citizens had made a commitment to show up at the city council meeting on the evening of the vote on the matter. It was an amazing show of support, and Trulio was hopeful that she could save the owls.

It was not to be, however. The afternoon before the scheduled vote, Trulio received a phone call from another activist saying that the developer had gone out to the site and plowed up most of the acreage of concern. Although destroying active owl burrows is illegal, prosecution depends on finding a dead bird. This proved impossible. To make matters worse, the city council voted in favor of the developer later that evening.

Clearly the current legal status of the burrowing owl in California, as well as in most other urban and agricultural areas, will not save them. Lobbying for the more stringent "endangered" classification in more regions is, of course, an option, but many feel that such efforts would stir up the public's ill will. The best approach, say some, is to raise the awareness of city planners and individual landowners, emphasizing that open natural spaces within our communities are more than just a home for wildlife. They provide much-needed breathing room and recreational space for our own species as well, adding to our quality of life, benefiting our health, and (this may be the clincher) potentially raising overall land values. The burrowing owl and other species at risk in our communities can serve as speed bumps to slow the pace of urban sprawl and curb excessive tillage and use of pesticides by farmers, thereby protecting us from ourselves. There is the potential for synergy between individual efforts to save the burrowing owl, the prairie dog, and the black-footed ferret. If realized, this synergy may ultimately lead to better protection for all three species and preservation of the vast grassland ecosystems of the American West.

9

THE GOOD EARTH

*And he stooped sometimes and gathered some of
the earth up in his hand and he sat and held it
thus, and it seemed full of life between his fingers.
And he was content, holding it thus.*

—PEARL BUCK, *THE GOOD EARTH* (1935)

*Man stalks across the landscape, and desert
follows his footsteps.*

—HERODOTUS (FIFTH CENTURY B.C.)

OUR CLEVER SPECIES HAS NOT YET LEARNED how to tread lightly on this good Earth. And it is from this good Earth, the soil to be
more precise, that we obtain more than 97 percent of our food
needs. Our attempt to conquer nature in the Great Plains during
the late nineteenth and early twentieth centuries had many environmental repercussions besides the near-eradication of prairie
dogs and other animal species. Our disregard for the need to adopt
soil conservation practices in the region led to the period of severe
soil erosion in the 1930s that became known as the Dust Bowl. Dust
storms as spectacular as those witnessed during the Dust Bowl era
are rare events today, but the rate of soil erosion in the United States
continues to exceed the rate of soil formation by at least a factor of
ten, and the situation is much worse in many other parts of the
world.

Many of our activities, in addition to those causing erosion, have
an insidious and global impact on the living soil resource and our
future food security. We are dumping toxic wastes onto the land at an
unprecedented rate. Acid rain and climate change, brought about by
our pollution of the atmosphere, can directly affect life in the soil,

and thereby affect nutrient cycling and other processes important to all life on the planet.

In this chapter, I focus on some of these issues—soil erosion, toxic waste, and climate change—from the perspective of the connection between humans and life in the underground. Like many scientists, I am concerned about the long-term consequences of our actions. Today we have a much better understanding of what those consequences are than the settlers of the Great Plains did a century ago. The question is, how will we use this knowledge? For the sake of short-term convenience, some would have us ignore it. But as Edward O. Wilson, the Pulitzer Prize–winning ecologist, put it: "It is a mistake to dismiss a worried ecologist or a worried doctor as an alarmist." The challenge is to not wallow in hand-wringing and finger-pointing, but rather to use our knowledge to make better decisions for our future, and maybe even clean up a few of our past mistakes.

THE WIND WAS ALREADY BLOWING UP DUST as residents of western Kansas awoke on March 15, 1935. By now they were used to this; they were near the center of the Dust Bowl, and spring dust storms had become routine. But as the day progressed, the skies grew very dark, and people moved about nervously as it became apparent that they were in for one of the severe "black blizzards." By noon, the sun, when it could be seen at all, had taken on an ominous violet-green hue. Cars traveled with their headlights on until visibility became so bad that traffic on city streets and highways came to a standstill. Some motorists waited out the storm, while others panicked and abandoned their vehicles to follow fence lines or curbs, desperately seeking shelter in the nearest home. In the rural areas, cattle huddled against the fiercely swirling dust as they would against wind-driven snow. A nine-year-old boy became stranded and confused as he struggled for air in the fierce winds. He was found the next morning, traumatized and tangled in the barbed wire of a fence he had climbed to keep his head above the gathering dust. One hundred miles away, near Hays, Kansas, a seven-year-old boy had not been so lucky. A search party eventually found his lifeless body buried beneath a mound of talcum-like dust.

The drought that triggered Dust Bowl tragedies such as these began in 1932, peaked in 1935, and continued intermittently until the

APPROACHING DUST STORM IN MIDDLE WEST.   #23- CONARD

FIGURE 9.1 *Tons of topsoil were eroded from farmlands during the "black blizzards" that were common on the Great Plains in the Dust Bowl era. Courtesy of Kansas State Historical Society.*

return of a wet cycle in 1940. The center of the Dust Bowl was the Texas and Oklahoma panhandle region. Large sections of northeastern New Mexico, southeastern Colorado, and western Kansas were also affected. Some of the storms swirled far beyond this region, however, reaching the East Coast and going hundreds of miles out into the Atlantic Ocean.

The southern Great Plains had historically been prone to periodic drought cycles and strong spring winds, but because the native deep-rooted grasses of the region held the light, sandy soils in place, catastrophic wind erosion had usually been avoided. This changed during the period of rapid agricultural expansion between 1900 and 1930. The "sodbuster" farmers ripped up much of the native grasslands to plant annual cash crops, primarily wheat. The wheat and other cash crops did not hold the soil as well as the native grass sod, and large tracts of land were often left completely barren of all vegetation for parts of the year. Overgrazing by livestock also degraded grasslands during this period.

Many of the landowners of the southern Great Plains were "suit-case farmers" who lived in urban areas far away and leased small plots to relatively poor tenants, who had no vested interest in the land. Owing to escalating land prices, the percentage of tenant farmers increased to nearly 40 percent by 1930. With distant landowners and so many tenant farmers, few people were taking a long-term view toward soil conservation.

Moreover, many farmers overestimated the resilience of the land, including one in Guymon, Oklahoma, quoted as saying, "Let the wind blow the topsoil away. We can plow up more. . . . You just can't seriously hurt this land out in the Panhandle." Of course, this was not the case. The nutrient- and microbe-rich topsoils of the Great Plains, just like topsoils everywhere, require thousands of years to form but can degrade quickly with mismanagement. It often took just five to six years after sodbusting for land in the Panhandle to turn from highly productive to useless for commercial wheat farming. The wealthy landowners could afford to abandon these sites and move on to others, leaving the abandoned fields devoid of vegetation and highly susceptible to further erosion. The introduction of large tractors driven by steam and combustion engines made matters worse because they led to excessive plowing and pulverization of the soil. The misuse of these marvels of technology, referred to as "snubnosed monsters" by John Steinbeck in his novel *The Grapes of Wrath*, played a major role in the Dust Bowl tragedy.

Unusually high rainfall between 1900 and 1930 had made farming much easier and was one reason for the rapid settlement and transformation of the Great Plains landscape at that time. For example, the average annual rainfall recorded in Cimarron County, Oklahoma, from 1900 to 1930 was 20 inches (50 centimeters), and as high as 28 inches from 1914 to 1923. In contrast, between 1934 and 1939 the region underwent a drought cycle with annual rainfall of less than 14 inches, and in some areas less than 8 inches. During the 1930s, an entire generation of farmers that had grown up during an unusual wet cycle and a period of little threat from wind erosion was suddenly faced with severe drought and the repercussions of decades of abuse of the soil resource.

The Kansas Academy of Science developed a complex system for describing the Dust Bowl storms, with a number of basic classifications and several "species" within each class. But to the local citizens and journalists, the storms were all "dusters," and the worst of these were the "black blizzards": Localized events in which the dust boiled along close to the surface at speeds of thirty to sixty miles per hour or more, reducing visibility to near zero, covering fences, damaging homes and property, suffocating animals, and even, on occasion, suffocating people.

As terrifying as the localized black blizzards were to the residents of the Great Plains, a different kind of dust storm actually caused more wide-scale land degradation. Occasionally triggered by westerly low-pressure weather systems, these storms would sweep tons of topsoil from a wide geographic area several thousand feet up into the air, where the billowing red-brown clouds of dust could be captured by the jet stream and carried across the continental United States. Although the dust created at the surface by these low-pressure system storms was less dense than occurred with the black blizzards, they still darkened skies with a thick haze that irritated the eyes, nose, and throat and caused respiratory distress.

On May 12, 1934, the *New York Times* reported that Manhattan "was obscured in a half-light similar to the light cast by the sun in a partial eclipse. . . and much of the dust seemed to have lodged itself in the eyes and throats of weeping and coughing New Yorkers." The same storm cast a gloomy shadow over Washington, D.C. Easterners did not appreciate their first full taste of Great Plains grit, but it put the Dust Bowl crisis on the national political agenda. In this one event alone, it was estimated that at least 300 million tons of precious topsoil billowed up from the farmlands of the southern Great Plains and was deposited over the eastern half of the nation.

During the spring months of 1932 to 1939, dusters of all types were so frequent that residents of the Dust Bowl region often kept the windows of their homes sealed with tape and covered with wet sheets to filter the air. The dirt still got in. It became habit in many households to set the dinner table with the plates upside down until the meal was ready to be served. By the mid-1930s, some school districts had begun to close a month early to minimize the risk to students of being caught in a storm. Poor air quality became a chronic problem. Doctors reported a dramatic increase in respiratory illnesses, and several deaths were attributed to "dust pneumonia." It became commonplace for people to wear a handkerchief across their nose and mouth when they stepped outside. During a legislative debate on the crisis in the Texas state legislature in 1935, most of the senators were said to be wearing surgical masks.

There was much finger-pointing during the Dust Bowl era as everyone looked for someone or something to blame. The drought from 1934 to 1937 was the worst on record, and it certainly was a major factor in creating the Dust Bowl. But elimination of much of the native grassland for agricultural purposes, excessive plowing, and overgrazing had undoubtedly set the stage for this environmental tragedy. By the late 1930s, the role of human activities in bringing on this catastrophe was apparent to nearly everyone, and new soil conservation strategies began to be implemented in the region.

One good thing to come out of the Dust Bowl was the Soil Erosion Act passed by Congress in 1936, which committed the federal government to a massive soil conservation effort. Six million of thirty-two million cultivated acres in the southern Great Plains identified as being at high risk for erosion were taken out of production and converted to permanent grassland to stabilize the soils. Through education and subsidy payments, farmers were encouraged to adopt practices that would slow the rate of soil degradation, such as reducing the frequency of plowing and utilizing tillage equipment designed to minimize damage to soil structure. Contour plowing and terracing were encouraged in hilly regions.

This effort spilled out beyond the boundaries of the Great Plains. The federal Soil Conservation Service and state-created conservation districts raised awareness of the threat of soil erosion and introduced new approaches to land management throughout the nation. Although erosion remains a critical problem in the United States today, the rate of soil erosion during the latter half of the twentieth century declined as a direct result of the conservation plans implemented after the Dust Bowl.

Despite the lessons learned during the Dust Bowl, no nation in the world is managing its agricultural soils in a sustainable way as we enter the twenty-first century. The current estimated rate of soil loss from U.S. and European croplands due to erosion is about four to six tons per acre (nine to fifteen metric tons per hectare) per year. That's enough soil to fill about a dozen small pickup trucks, and a rate of soil loss more than ten times the rate at which new topsoil can be formed by natural processes. Although erosion occurs in forests and other natural ecosystems as well, the rate is much slower and more sustainable than it is with farm soils, which are repeatedly plowed and periodically left without a protective cover of vegetation. On lands that are fallowed (left barren of any vegetation), several tons of topsoil can be washed from a one-acre field in a single severe rainstorm. In terms of depth of soil lost, the average erosion rate for the United States and Europe represents just a fraction of an inch per year, a rate that is imperceptible with the naked eye. But soils form very slowly. It takes between two hundred and one thousand years to form one inch (two and a half centimeters) of new topsoil. The uppermost layer of topsoil, which is the first to be eroded, unfortunately also happens to be where we find most of the soil microbes essential to decomposition, nutrient cycling, and healthy crops, and the highest concentration of plant nutrients.

The erosion problem in the United States today is less spectacular than what was witnessed during the Dust Bowl era, but precisely because of this, it is a more insidious threat. Soil erosion is often put on the back burner of the national political agenda. Many of the negative impacts on soil quality and crop productivity are being masked in the short term by increased use of fertilizers and more frequent irrigation. But if it continues unabated, the current erosion pattern could prove more harmful to the environment and the economy in the long term than the ferocious dust storms of the 1930s.

Soil erosion in some other parts of the world, particularly the humid tropics, is nearing crisis levels. Many tropical developing countries are losing thousands of acres of arable land each year due to rainfall erosion in hilly regions that are being farmed without adequate attention to soil conservation. For example, along the mountain slopes of central Java, the Machako and Aberdare regions of Kenya, the lower Himalayas of India and Nepal, and the Andes Mountains of South America, erosion rates are often between thirteen and eighteen tons per acre per year, and rates as high as forty tons per acre per year and higher have been measured. One year's damage at these rates will take centuries to repair. Peasant farm families watch helplessly as their cropland becomes gouged with gullies, and the reduction in soil depth within a single rainy season is often shockingly visible. When eroded fields become unproductive, they are abandoned and new rain forest areas are clear-cut for agricultural purposes. Erosion is a leading cause of deforestation in the tropics. Compounding the problem further is the fact that the nations of the humid tropics are also where food shortages are most severe, population growth rates are highest, farmers are poorest, and the infrastructure for developing regional soil conservation strategies is weakest.

When erosion is severe and widespread, the economic impact is readily apparent. When crop production in the southern Great Plains declined by about 70 percent due to erosion between 1900 and 1940, the economic impact reverberated throughout the United States. Similarly dramatic effects are being experienced in parts of the tropics today.

When erosion is not extreme, effects on crop productivity can sometimes be masked in the short term by spending more on fertilizer, irrigation, and other inputs. Some analysts include such expenditures in their analysis of erosion costs while others do not, resulting in very different estimates of economic impact. Also, different approaches are used to estimate the amount of soil and nutrients lost from a region due to erosion. Some analysts rely on estimates of the amount of sediment washed into streams and rivers within a watershed. Others rely exclusively on mathematical models that estimate erosion from soil type, topography, and weather. Still others rely on measured soil loss data collected from individual fields, and extrapolate to the regional scale. The only consensus in the scientific community is that none of these methods is completely satisfactory. For now we are left with highly variable estimates. For example, recent estimates for annual costs of erosion to agriculture in the United States have ranged from $600 million to $27 billion.

The economic impact of erosion reaches far beyond the farm gate, of course, and this impact must also be assessed. More than half of the billions of tons of topsoil lost from cropland each year ends up as sediment that clogs our streams, rivers, and reservoirs. Some of this sediment contains pesticide residues and fertilizers that are harmful to fish and other aquatic wildlife. The United States spends more than $500 million each year to dredge sediments from waterways as a flooding prevention measure, but this strategy is only partially effective. The widespread flooding and subsequent economic losses in the midwestern United States in 1993 were attributed in part to erosion and the resulting sedimentation of the Mississippi and Missouri Rivers. Sediment accumulation in reservoirs reduces water storage capacity and electricity production, increases maintenance costs, and in some cases has contributed to earth dam failures. Other off-site costs of erosion include cleanup or repair of roadways, sewer systems, and basements, and abrasive wind damage to property. Estimates of total off-site costs of erosion vary depending on the methodology used to attain them, just as with the on-farm costs. In the United States alone, the off-site cost of erosion may be as high as $17 billion per year, according to one analysis.

Reducing soil erosion on farms to sustainable rates—that is, rates similar to those measured in natural grasslands and forests—is feasible. Severe soil erosion can usually be prevented by following just a few basic rules: Maintain as much vegetative cover on the land year-round as possible; plow as little as possible; do not farm sloping land (or if you must, use contour planting or terracing); and establish "windbreak" rows of tall annual crops or trees in areas prone to wind erosion.

The simple explanation for why farmers everywhere aren't following these straightforward soil conservation guidelines is that they often find it more profitable in the short term to ignore them. Many farmers cannot afford to set aside steeply sloping land, and they may not have the equipment or labor for establishing terraces. Maintaining vegetative cover year-round involves seeding and cultivating noncash "cover crops" for part of the year that bring in little or no revenue. From a soil conservation standpoint, it is best to leave all crop residue in the field after harvest, but peasant farmers in many parts of the world rely on crop residue, including the roots, for cooking fuel. From a soil conservation standpoint, a farmer should not enter a field with heavy farm machinery when it is wet because doing so severely compacts the soil, leading to greater susceptibility to runoff and erosion in the future. In the United States and many other regions, however, farmers frequently break this rule in order to harvest crops at the optimum stage of maturity or when market prices are highest.

Most governments have begun to recognize that preservation of soil resources must be a national priority. It is risky, and arguably unfair, to leave this responsibility entirely up to the farming community, particularly when the allure of higher short-term profits tempts farmers to ignore soil conservation practices. Since the Dust Bowl, many developed nations have slowed the pace of soil erosion with comprehensive conservation plans that include economic incentives for farmers. However, much more clearly needs to be done. In the recent past, the total global investment in agricultural and environmental research and development has amounted to less than 1 percent of the total spent on national defense. This disparity may have to change if we are to effectively tackle soil erosion, which is a threat to future food security everywhere.

ACCELERATED EROSION IS JUST ONE EXAMPLE of a negative impact of human activities on soil resources. Another is our poisoning of the soil habitat and groundwater supplies by careless dumping of toxic wastes onto the land. This is the outcome of an optimistic assumption, which we have maintained for the sake of convenience, that the healing and purifying powers of the soil are nearly infallible and will eventually make harmless whatever we put there.

It was perhaps in this spirit that the officials of the Hooker Chemical Corporation buried about 20,000 tons of chemicals, many of them known carcinogens, near their manufacturing facilities at Love Canal, New York, between 1942 and 1953. Nothing was said when schools and residences were later constructed on the site. Twenty years later, it became evident that a terrible mistake had been made. Love Canal residents were suffering from unusually high rates of cancer, birth defects, and other maladies. In 1978 the area was pronounced unfit for human habitation. Since then, we have discovered many other examples of toxins accumulating in soils, groundwaters, and sometimes the living tissues of plants, animals, and humans. We are now faced with thousands of toxic waste sites, polluted soils and water supplies, and a tremendous volume of nonrecyclable plastics.

Many soil microbiologists today are dedicating their careers to coming up with solutions to this problem. They are utilizing the latest information on subterranean life to develop methods for bioremediation of our polluted land areas and toxic sites. Sometimes relatively inexpensive methods are effective. For example, the growth and activity of beneficial de-toxifying microbes already present can sometimes be stimulated just by adding nitrogen or phosphorus fertilizers or by aerating the soil. "Seeding" a toxic site with beneficial microbes brought in from elsewhere or purchased from commercial manufacturers is another option. The annual sales of microbial bioremediation products currently exceeds $10 million, and the potential market may be as high as $200 million. Commercially available, petroleum-eating bacterial populations have been used to clean up both land and sea oil spills, such as the infamous *Exxon-Valdez* spill off the coast of Alaska.

Many of the microbiologists interested in bioremediation are in the business of searching the globe for new microbes to help us clean up our mistakes. The ideal microbes are those that either thrive on toxic compounds as a food source or inadvertently break the toxins down, and that can survive in the harsh environments in which toxins often occur. Many of those discovered that are proving useful are among the unusual extremophile bacteria and archaea discussed in chapters 2 and 3.

Some of the microbes used in bioremediation are not that exotic. During the past decade, scientists stumbled upon the fact that a relatively common group of soil fungi in the genus *Phanerochaete* excrete powerful enzymes that can break down some of our most troublesome toxins. These so-called white-rot fungi normally use their enzyme arsenal to decompose the rotting wood in forested areas. The same enzymes have proven amazingly effective at breaking down some of our most troublesome pollutants, such as dichlorodiphenyltrichloroethane (DDT) and other insecticides, polychlorinated biphenyls (PCBs), the explosive trinitrotoluene (TNT), and various types of plastics.

Toxic heavy metals, such as uranium, selenium, arsenic, and mercury, represent another category of pollutant that may soon be treatable by bioremediation. Soil microbes that can transform these metals into harmless products have been discovered. The various de-toxifying mechanisms of these microbes usually involve tying up the metals into compounds that make them harmless or less likely to be absorbed into the tissues of plants and animals.

Some of the toxins in our environment are synthetic organic compounds that have been created and released on the Earth for the first time by our species. These are often the most difficult to get rid of because the pace of human ingenuity has outpaced microbial evolution. When microbes with desired properties cannot be found, scientists may be able to genetically engineer new types for specific bioremediation purposes. Some successes have already been achieved, such as new strains of the bacteria *Pseudomonas putida* that can degrade TNT and the herbicide 2,4,5-T more completely than natural strains. Sometimes a microbe capable of breaking down a toxin already exists but is not useful because it cannot survive the high acidity or other harsh environmental conditions found at a toxic site. In this case, genetic engineering can be used to provide the beneficial microbe with the necessary tolerance to environmental stress.

Field-testing of genetically modified organisms specifically designed for toxic cleanup is under way. There are legitimate concerns that the use of such organisms could introduce new biological hazards even as the chemical hazard is alleviated. The majority of microbiologists working on bioremediation are convinced that these fears can be put to rest with appropriate field evaluation and regulatory oversight. Risk assessment and difficult choices will need to be made in the future because these new microbes may be our only hope of cleaning up some of our most hazardous toxic waste sites.

ONE MIGHT THINK THAT CREATURES of the underground would be sheltered from at least one form of our pollution, that which we put up into the atmosphere, but such is not the case. Even those soil species that are never directly exposed to changes in surface air quality are vulnerable to two potential consequences of atmospheric pollution—acid rain and climate change. Also, the subterranean world is very sensitive to the impact of toxic and greenhouse gases on Earth's plant life, which serves as its primary food source.

The acid rain story began as a mystery. During the 1970s, many of the spruce, fir, and other evergreen trees in the forests of central Europe were coming down with a strange malady. The symptoms often began with a yellowing of the needles, followed by needle drop, slowed growth, and sometimes premature death. Policymakers began to pay attention when it became clear that economically important timber trees, such as Norway spruce (*Picea abies*), were among those affected over a broad geographic range. A comprehensive study conducted by the German government in the 1980s concluded that 20 to 25 percent of European forests could be classified as moderately to severely damaged from "unknown causes."

Among the many hypotheses proposed to explain this peculiar forest decline were unusual weather patterns, plant disease, and nutrient deficiencies. Under scrutiny, however, none could fully explain the chronic, widespread nature of the problem. When spruce and fir trees in the northeastern United States began to show damage similar to that in European trees, another hypothesis was added to the list—

FIGURE 9.2 *Severe needle loss and general decline of Norway spruce trees in central Europe associated with acid rain effects on soil biological and chemical properties. Courtesy of John Aber, University of New Hampshire, from University of Bayreuth study.*

direct damage to the foliage of the plants by atmospheric pollutants. Both regions had relatively poor air quality due to high population densities and industrial activity. But after several years of research, this hypothesis also was rejected, because the symptoms did not resemble classic pollution damage and the severity of the tree deterioration was not well correlated to concentrations of atmospheric pollutants.

The air pollution hypothesis led the scientists in the right direction, however. What all of the regions with forest decline had in common was exposure to acid rain, which occurs when gaseous air pollutants—usually sulfate, nitrous oxide, and/or ammonia gas—dissolve into airborne water droplets that fall to Earth as rain. The major sources of sulfate are coal-burning power plants and auto exhaust. Nitrous oxide and ammonia gases are released in trace quantities from most soils, but as mentioned earlier in the discussion of the nitrogen cycle, the amount that escapes from soils increases near regions of intensive agriculture, where there are high inputs of nitrogen fertilizers or manure.

It could be shown that the soils in the regions with forest decline were becoming acidified and nitrogen-enriched owing to acid rain. This process was changing soil chemistry and microbial activity, with adverse effects on the availability of some plant nutrients. It is well known that calcium (the fifth most abundant element in trees), magnesium, and potassium often become less available to plant roots in acid soils. At the same time, some potentially toxic elements, such as aluminum, become more abundant. Classic symptoms of calcium deficiency and aluminum toxicity were observed in many of the regions of forest decline.

The nitrogen in acid rain, rather than serving as a beneficial "fertilizer," appears to increase the severity of nutrient imbalances in some evergreen tree species in areas where forest decline has been observed. The nitrogen initially stimulates more growth, but this increases the demand for calcium and other soil nutrients that have become less available because of increases in soil acidity. The end result is more severe nutrient deficiencies and weak plants that are more vulnerable to disease and environmental stress.

The amount of nitrogen deposited by acid rain can be substantial. Annual rates of deposition of atmospheric nitrogen to the forests of central Europe and the northeastern United States are often in the range of thirty to forty pounds per acre (thirty-five to forty-eight kilograms per hectare), and rates as high as ninety pounds per acre have been measured in forests that border urban centers.

Research on another front has revealed that the effects of acid rain on soil biology as well as soil chemistry may be involved in the observed forest decline. In a comprehensive, multiyear analysis, Eef Arnolds of the Wijster Biological Station in the Netherlands documented a dramatic decline in populations of soil mycorrhizal fungi throughout central Europe during the 1970s and 1980s. As we know, trees are particularly dependent on their symbiotic relationship with these fungi to assist their roots in the acquisition of water and nutrients and to help them survive periods of drought and nutrient stress. Arnolds's study detected the most severe decline in mycorrhizal populations among the spruce and fir trees that showed the most severe health problems and in regions where acid rain and nitrogen deposition levels were highest. In many cases, Arnolds was able to document that the decline in mycorrhizae preceded the decline in tree health. Separate experiments have shown that the growth and activity of these mycorrhizal species are inhibited by increases in soil nitrogen and acidity. It appears that acid rain has direct negative effects on both members of the symbiosis—the trees and the root fungi—and that these negative effects are exacerbated by the fact that the health of each is dependent on the well-being of the other.

As we enter the twenty-first century, the emission of sulfates has been almost cut in half in both the United States and Europe owing to new regulations and the implementation of cleaner methods of burning fossil fuels at power plants. The nitrogen-based air pollutants are proving to be much harder to monitor and control. A challenge for the twenty-first century will be coming up with practical solutions to this problem. As discussed earlier, research is under way to develop crops and farming methods that improve nitrogen-use efficiency. Simply restricting nitrogen fertilizer use without completing this first step could not only have negative economic impacts and jeopardize food supply but result in a demand for more land to meet food needs, thus accelerating agricultural expansion into natural ecosystems.

NITROUS OXIDE, IN ADDITION to being a causative agent of acid rain, is also one of several so-called greenhouse gases and reduces the concentration of stratospheric ozone (Earth's UV radiation shield). Soil microbial activity, as affected by human applications of nitrogen, is a major source of this gas. Soil microbes are also very involved as both a source (producer) and a "sink" (absorber) of several other important greenhouse gases, most notably methane and carbon dioxide.

The relative contributions of excess nitrous oxide, methane, and carbon dioxide emissions to total "global warming potential" are estimated to be about 5, 15, and 60 percent, respectively. These values are calculated from estimates of gas concentrations and the per-molecule potency of each gas in trapping heat and causing warming. All three of these gases are steadily on the rise (figure 9.3), with humans as the driving force and soil microbes often as the unwitting mediators of change. Because soil organisms are so important as both sources and sinks of greenhouse gases, how we choose to manage the soil habitat will play a major role in determining the magnitude of human-induced climate change. The extent to which our activities do ultimately alter the climate will affect life in the underground—benefiting some organisms, but possibly driving others to extinction.

FOR MANY THOUSANDS OF YEARS PRIOR to the twentieth century, atmospheric methane concentrations were relatively constant because the sources and sinks of the gas were roughly in balance. A major natural source of methane is production of the gas by the unusual microbes called methanogens (see chapter 3), the ancient, bacteria-like creatures of the Archaea domain that thrive in low-oxygen environments in the soil and elsewhere. Well over half of the methane pumped into the atmosphere each year is the waste by-product of the activity of these microbes.

In the past several decades, human land management activities have inadvertently created more environments that favor the activity of the methanogens, and the rate of this activity has exceeded the rate of removal of methane from the atmosphere by natural chemical and biological processes. Much of the current rate of increase (figure 9.3) is attributed to the increased acreage of flooded agriculture (primarily rice), whose low-oxygen soil conditions stimulate methanogen activity. Landfills are another expanding source of methane production. Reducing the acreage of flooded agriculture and changing landfill management operations could help reduce methane production, but these measures may not always be economically feasible. Also, we still do not have a complete picture of all of the sources and sinks for this greenhouse gas, so the benefit of such actions is hard to predict. As we learn more, the goal will be to minimize the activity of methane-producing soil microbes while encouraging the activity of some others that absorb methane and transform it into harmless compounds.

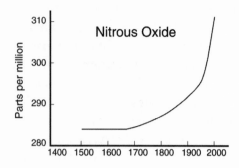

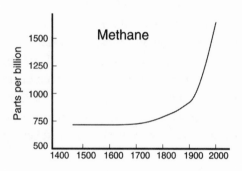

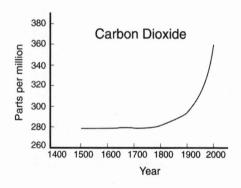

FIGURE 9.3 *Soil organisms play a major role as both sources and sinks of the three important greenhouse gases that have risen in atmospheric concentrations since the industrial revolution. Adapted from R. Lal et al.,* The Potential of U.S. Cropland to Sequester Carbon and Mitigate the Greenhouse Effect *(Chelsea, Mich.: Ann Arbor Press, 1998).*

Soil microbes also have a significant impact on atmospheric concentrations of carbon dioxide, the most important of the greenhouse gases. Historically, the primary sources of this gas have been the respiration of soil microbes and other creatures on the planet (including ourselves) and the occasional volcanic eruption. The primary natural sinks have been green plants and photosynthetic microbes, as well as chemical absorption in the oceans. Until we began burning oil, coal, and other fossil fuels, these natural sources and sinks were roughly in balance, and atmospheric carbon dioxide concentrations had been nearly constant for tens of thousands of years. Prior to the introduction of fossil fuel burning, this natural cycle was essentially just moving the same carbon back and forth between the atmosphere (as carbon dioxide) and the soil and biosphere (as organic carbon). When we pump fossil fuels up from the deep Earth and burn them, we are adding to the system new carbon that has been buried for millions of years and essentially "out of the loop." It is largely owing to this human addition of new carbon that atmospheric carbon dioxide is rising at an alarming rate (figure 9.3) and is expected to double within this century.

Plants are directly affected by changes in atmospheric carbon dioxide because they take up this gas to produce sugars during photosynthesis. Whether or not there is a concomitant climate change, we can be relatively certain that the anticipated doubling of carbon dioxide within this century will change the mix of plant species in many areas and alter the quantity and nutritional quality of plant biomass that most subterranean creatures rely on as their primary source of food. Research has begun, but at this time we still know very little about how this will affect decomposition, nutrient cycling, and other soil processes important to our own survival. The effects could be profound, particularly if accompanied by a significant change in climate.

Scientists have begun to evaluate land management options that might take some of the excess carbon dioxide out of circulation and sequester it in the soil. There are two components to this strategy. The first is to maintain actively growing plants on as much of Earth's land surface as possible so that carbon dioxide uptake through photosynthesis is maximized. The second is to sequester in the soil as much of the carbon-rich biomass of dead plants as possible. The goal is to gradually increase the proportion of carbon that is in the form of slowly decomposing organic matter in the soil and decrease the proportion that is in the form of a greenhouse gas in the atmosphere.

In practice, the land management techniques that will increase carbon storage in soils are the same ones that prevent soil erosion and maintain crop productivity. It is estimated that when soils are eroded, as much as 20 percent of their carbon is converted to carbon dioxide and lost to the atmosphere. Erosion breaks up small soil clods, thus exposing more of the organic matter to oxygen and stimulating rapid and complete decomposition by aerobic soil microbes. Maintaining year-round vegetative cover, which slows erosion by holding the soil in place during periods of heavy rainfall or wind, has the added benefit of increasing the quantity of photosynthetic carbon dioxide uptake by plants. By minimizing plowing, a soil structure can be maintained that is more resistant to erosion; this technique also reduces aeration of the soil, thereby slowing complete decomposition and carbon dioxide loss from soils. All of these measures also tend to boost crop productivity. This is an opportunity for land managers to improve the soil resource upon which their livelihood depends while making a positive contribution to the environment on several fronts.

But will planting more trees, encouraging farmers to maintain vegetative cover, and increasing soil organic matter save us from global warming? I and others who have done the math on this question have concluded that the impact will be modest at best. We are pumping so much new carbon into the atmosphere each year in our burning of fossil fuels that this approach by itself is not a solution. Even if we could keep up initially, within a few decades we would run out of new land on which to plant trees and would saturate the capacity of the soil to store more carbon. This would be the outcome even if our optimistic assumptions about a carbon dioxide "fertilization effect"—a stimulation of the growth of photosynthetic plants—proved well founded.

Global warming will work against us in our efforts to store more carbon in soil because warmer temperatures stimulate the activity and carbon dioxide emission of most decomposing soil organisms. Nevertheless, the approaches to land management that maximize soil carbon storage are being embraced by farmers and environmentalists alike because they represent a "win-win" strategy—the economic and environmental benefits are substantial even if the impact on climate change is small.

I WAS BORN INTO a world with two and a half billion people. It took our species several thousand years to attain that population level. Within just the past fifty years, we have more than doubled it. There are now more than six billion of us sharing the planet. With each passing year, more and more of us are consuming more and more resources on a per-person basis. We all are players in a grand global experiment in progress. How will our voracious appetite for resources and our careless attitude toward the environment affect the Earth habitat that we depend on for our survival and quality of life? It is a social experiment as well: Now that we are beginning to recognize the magnitude of our deleterious effect on the environment, how will we respond as individuals and as a society?

The human footprint on the living soil resource already looms large. Erosion and the reckless dumping of toxic waste have caused severe land degradation in many regions. We have evidence that acid rain has jeopardized the symbiotic mycorrhizal root fungi in the forests of Europe and the northeastern United States. Scientists are relatively certain that atmospheric carbon dioxide concentrations will double within this century, and that this change by itself will significantly affect soil organisms and soil processes. If the rise in carbon dioxide and other greenhouse gases leads to a warmer planet, as most climatologists predict, many soil functions important to humans could be dramatically affected.

No matter what kind of havoc we wreak on the environment, most species of the underground are so tough, and able to evolve so quickly, that they will outlast us. Nevertheless, our activities are probably already inhibiting the function of many soil species, and driving a few toward extinction. With as many as ten thousand distinct species in a pinch of soil, can losing a few really matter? We cannot answer this with any certainty because so few have been studied. Some may be irreplaceable because they possess unique enzymes that control key steps in a vital aspect of nutrient recycling. The vast majority are probably playing backup to each other, with some overlap of their ecological roles. But a species that appears redundant and expendable under one set of conditions may become essential when the environment changes. The strongest argument for preventing the loss of even one species is that we cannot predict what characteristics might be of value to us in the future. The species we allow to go extinct today might be the only one that can produce a unique antibiotic to fight a disease that emerges twenty years from now, or it may be uniquely able to break down a new synthetic toxin that we haven't even invented yet.

The commercial potential of "bioprospecting" the soils and deep Earth for new organisms with desirable properties is one incentive for protecting subterranean biodiversity and soil habitats. Another incentive is the fact that by the middle of this century we will have at least ten billion mouths to feed, four billion more than today, and this population will put tremendous pressure on soil resources. Unambiguous economic incentives such as these have already caused some changes in our behavior.

In just the past decade, farmers in many developed nations have substantially changed their approach to land management. They have come to realize that synthetic chemicals are not the solution to every problem, and that overreliance on synthetic fertilizers—as opposed to soil-building crop rotation strategies—has contributed to a depletion of soil organic matter and lowered productivity. Modern farming is knowledge- not chemical-based, and much of the new knowledge being applied to food production involves management techniques that conserve soil resources and take full advantage of beneficial soil organisms. Although erosion continues to be a serious worldwide problem, we have at least demonstrated that with a concerted effort the trend can be turned in the right direction.

We have seen improvements in other industries besides agriculture. Sulfate emissions by power plants have been nearly cut in half in the United States and Europe, and at a much lower cost than was originally anticipated. Although we have not been as successful at reducing nitrous oxide and some other air pollutants, poor air quality and acid rain are problems on the mend in many areas.

Our response to encouraging news such is this can be one of "conditional optimism," to borrow a phrase that Daniel Hillel uses in his book *Out of the Earth*. Conditional optimism moves us beyond hand-wringing and pessimistic despair, but is more realistic than the "pathological optimism" of those who naively assume that science and technology will come up with a quick fix to avert the environmental train wreck we are headed toward. Conditional optimism leads us on a pro-active path that requires a shift in behavior at both the individual and societal levels. It recognizes that applying our scientific know-how to the task of cleaning up damage after it has been done is not enough. Our knowledge must also be used to develop "win-win" strategies that minimize the environmental impacts of our activities while allowing us to improve our quality of life. We have already seen positive signs that we can accomplish this if we put our minds to it. Such action, combined with a tempering of our aggressive instinct to "conquer nature," will be necessary if we are to protect the living soil resource for future generations.

Epilogue

A song of the rolling earth, and of words according,
Were you thinking those were the words, those
*   upright lines? those curves, angles, dots?*
No, those are not the words, the substantial words
*   are in the ground and sea,*
They are in the air, they are in you.

—WALT WHITMAN, LEAVES OF GRASS (1855)

THE NINETEENTH-CENTURY POET WALT WHITMAN was sharply
critical of scientists when their reductionist approach to the study of
nature became obsessive, and they missed the forest for the trees. He
believed that the imaginative faculty of poets and artists was essential
to the "vivification" of scientific facts and that, until this occurred,
our concept of reality would remain incomplete. With this in mind,
whenever I have imagined an audience for this book, the naturalist
poets are always there, often gazing over my shoulder, the first to see
what I have come up with.

Poets and scientists have at least one thing in common—a sense of
wonder for the world around them. When Whitman wrote the classic *Leaves of Grass*, almost nothing was known of soil biology and
chemistry. Yet some of his words reveal an intuitive sense of the
importance of the underground to our species and all others on the
planet. Like this passage:

What chemistry!. . .
That when I recline on the grass I do not catch any disease,
Though probably every spear of grass rises out of
   what was once a catching disease.
Now I am terrified at the Earth, it is that calm and patient,
It grows such sweet things out of such corruptions,. . .
It gives such divine materials to men,
   and accepts such leavings from them at last.

What poetry might Walt Whitman have been inspired to write had he known about the vast populations of extremophile microbes living in the deep Earth? Or the underground fungal network connecting land plants to each other? Or the antibiotics from the soil capable of curing our most deadly diseases? More important, what insight might he bring to our discussion of twenty-first-century environmental issues, such as the judicious use of soil microbes in biotechnology, and our competition for space and resources with other species, such as the prairie dog, black-footed ferret, and burrowing owl?

Since Whitman is not here to help us transcend the chasm between the science of the underground and the true nature of things, it will be up to others to take on this responsibility. There is a bit of the poet in each of us, and ultimately that is what we must rely on to form our personal sense of the natural world and our place in it. Perhaps learning about some of the fascinating new discoveries described in this book has already expanded your appreciation for the planet we inhabit. This knowledge has certainly had such an impact on me.

Of course, the "facts" are constantly changing. In the process of writing this book, I have been continuously revising chapters written earlier. It has been difficult keeping up. But now my job is complete, and it is up to you, the reader. Rather than view the tales presented here as a collection of facts, they should be viewed as an attempt to whet your appetite and make you alert to new tales from the underground. New discoveries, and theories to explain them, will certainly be forthcoming, and from many sources. It is my hope that as more of us become aware of the life beneath our feet, and its relevance to our well-being, we will be inclined to work together to maintain the biological integrity of the underground, and preserve some of what we find there for future generations.

Notes and References

Introduction

2: For a more comprehensive review of our current state of knowledge regarding **soil biodiversity,** see D. Wall and J. Moore, "Interactions Underground," *BioScience* 49(2, 1999): 107–17; and L. Brussaard et al., "Biodiversity and Ecosystem Functioning in Soil," *Ambio* 26(1997): 563–70.

2: **Leonardo da Vinci** is quoted in Daniel Hillel, *Out of the Earth* (Berkeley: University of California Press, 1991), p. 3.

4: The status of **subsurface biology research** has recently been assessed by the Committee on Soil and Sediment Biodiversity and Ecosystem Functioning, led by Dr. Diana H. Wall at Colorado State University. This is part of a larger international effort called SCOPE (Scientific Committee on Problems in the Environment). Their work is described in *BioScience* 49(2, 1999): 107–52, and *BioScience* 50 (12, 2000): 1043-1120.

5: Two excellent general **soil ecology texts** are D. Coleman and D. A. Crossley Jr., *Fundamentals of Soil Ecology* (San Diego: Academic Press, 1996), and Martin Wood, *Environmental Soil Biology* (London: Chapman and Hall, 1995).

6: It is a little dated now, but for a general soil science text, with information about **soil profiles** and much more, I recommend Hans Jenny, *The Soil Resource: Origin and Behavior* (New York: Springer-Verlag, 1980). For a much more lyrical treatment of soils *per se*, there is William Bryant Logan, *Dirt: The Ecstatic Skin of the Earth* (New York: Riverhead Books, 1995), which is based in part on Logan's travels and interviews with Jenny. The USDA *National Soil Survey Handbook* website (www.statlab.iastate.edu/soils/nssh) maintains the most recent taxonomic classifications for soils.

11: To learn more about the amazing **water bears,** see Stephen Jay Gould, "Of Tongue Worms, Velvet Worms, and Water Bears," *Natural History* 104(1, 1995): 6–15. Their ability to survive for up to one hundred years in a near-deathlike state is described in J. Crowe and A. Cooper Jr., "Cryptobiosis," *Scientific American* 225(6, 1971): 30–36.

Chapter 1: Origins

17: For a highly readable account of the evolution of our solar system and Earth in relation to the origin of life, see the special "Life in the Universe" issue of *Scientific American* 271(4, 1994): 44–91. This subject matter is also well explained in W. Schlesinger, *Biogeochemistry: An Analysis of Global Change* (San Diego: Academic Press, 1997), pp. 15–126.

18: Unique aspects of clays and other secondary minerals are explained in most geology and soil science textbooks. Some clay minerals were crystallized not on Earth but elsewhere, then delivered to our planet in meteorites that collided with our planet when it was first forming. For a detailed description of clay formation in relation to the origin of life, see A. G. Cairns-Smith and H. Hartman, eds., *Clay Minerals and the Origin of Life* (Cambridge: Cambridge University Press, 1986).

19: The underground as the safest place for early life forms is discussed in N. Pace, "Origin of Life—Facing up to the Physical Setting," *Cell* 65(1991): 531–33; and T. Stevens, "Subsurface Microbiology and the Evolution of the Biosphere," in *The Microbiology of the Terrestrial Deep Subsurface,* edited by P. Amy and D. Haldeman (New York: CRC Lewis, 1997), pp. 205–23.

20: Charles Darwin's letter referring to the origin-of-life question as *ultra vires* is quoted from *Evolution from Molecules to Men,* edited by D. Bendall (Cambridge: Cambridge University Press, 1983), p. 128.

20: Thomas Huxley's essay, "On the Physical Basis of Life," can be found in T. Huxley, *Lay Sermons, Addresses, and Reviews* (New York: D. Appleton and Co., 1871), pp. 120–46.

21: William Fowler's Nobel lecture, "The Quest for the Origin of the Elements," was published in *Science* 225(1984): 922–35.

22: Christian de Duve's comment on the "pathway to life" is from his book *Vital Dust: Life as Cosmic Imperative* (New York: Basic Books, 1995), p. 24.

23: The original paper by Stanley Miller was "Production of Amino Acids Under Possible Primitive Earth Conditions," *Science* 117(1953): 528–29. Among the many excellent reviews of origin-of-life research, the two that I turned to most often were P. Davies, *The Fifth Miracle: The Search for the Origin and Meaning of Life* (New York: Simon and Schuster, 1999); and L. Orgel, "The Origin of Life on Earth," *Scientific American* 271(4, 1994): 76–83. Also helpful were de Duve, *Vital Dust,* and C. Wills and J. Bada, *The Spark of Life* (Cambridge, Mass.: Perseus Publishing, 2000).

25: The quote by Paul Davies is from *The Fifth Miracle,* p. 91.

26: The quote by Chief Seattle is from J. Campbell, *The Power of Myth* (New York: Doubleday, 1987), p. 34.

26: The molecular structure of clay crystals, their unusually high surface area, and their chemical reactivity are described in most soil science texts, such as in Jenny, *The Soil Resource.* The unique properties of clays are discussed in less technical fashion in Logan, *Dirt.*

28: The theory of clays acting as enzymes and the template for the synthesis of complex organic molecules is covered in detail in Cairns-Smith and Hartman, *Clay Minerals and the Origin of Life,* pp. 1–10, 130–51.

28: John Desmond Bernal speculates on the importance of clays in his book *The Physical Basis of Life* (London: Routledge and Kegan Paul, 1951).

28: The binding of nucleotides to clay surfaces was reported in James Lawless et al., "PH Profile of the Adsorption of Nucleotides onto Montmorillonite," *Origins of Life* 15(1985): 77–88. Research involving the actual sequencing of nucleotides by clays was summarized in J. P. Ferris et al., "Synthesis of Long Prebiotic Oligomers on Mineral Surfaces," *Nature* 381(1996): 59–61. The research by Lelia Coyne on the energy storage property of clays was reported by A. G. Cairns-Smith in "The First Organisms," *Scientific American* 252(6, 1985): 90–100.

29: Erwin Schrödinger's prediction that genes would be aperiodic crystals is found in E. Schrödinger, *What Is Life?* (Cambridge: Cambridge University Press, 1944), p. 64.

30: A popular version of the "clay gene" hypothesis can be found in A. G. Cairns-Smith, "The First Organisms," *Scientific American* 252(6, 1985): 90–100.

32: The role of clays in the origin of life remains controversial, but the basic concept that mineral surfaces of some kind played a crucial role in catalyzing Earth's first biosynthesis is becoming widely accepted. Of the alternatives to clay, the simple iron-sulfur compound pyrite is a current favorite. The paper that triggered the interest in pyrite was Günter Wächtershäuser, "Before Enzymes and Templates: Theory of Surface Metabolism," *Microbiological Reviews* 52(1988): 452–84. Recent experiments have verified that iron-sulfur compounds are stable and can perform their lifelike synthesis reactions at high temperature and pressure conditions similar to those in deep Earth and hydrothermal vents; see G. D. Cody et al., "Primordial Carbonylated Iron-Sulfur Compounds and the Synthesis of Pyruvate," *Science* 289(2000): 1337–40.

Chapter 2: The Habitable Zone

35: The phrase "deep, hot biosphere" to describe the microbial habitat of deep Earth was first used by Thomas Gold in "The Deep, Hot Biosphere," *Proceedings of the National Academy of Sciences USA* 89(1992): 6045–49.

35: For background popular science reading on subsurface extremophiles, see S. J. Gould, "Microcosmos," *Natural History* 105(3, 1996): 21–68; and W. Hively, "Life Beyond Boiling," *Discover* 14(5, 1993): 87–91. More comprehensive and technical treatments of the subject can be found in

Microbiology of Extreme Environments, edited by C. Edwards (Milton Keynes, Eng.: Open University Press, 1990); and *The Microbiology of the Terrestrial Deep Subsurface,* edited by P. S. Amy and D. Haldeman (New York: CRC Lewis, 1997). Note that while I am using "extremophile" here to refer to organisms that survive at extremely high temperatures and pressures, it can also refer to organisms found in other types of extreme environments, such as icy-cold arctic waters or toxic or acidic soils.

36: A very readable summary of how subsurface biology research on Earth has expanded our notion of the **habitable zone** within the universe is given in G. Vogel, "Expanding the Habitable Zone," *Science* 286(1999): 70–71.

36: The description of the journey down into the **East Driefontein gold mine** was based on an interview with Dr. William Ghiorse, one of the scientists involved, and the article by K. Krajick, "Hell and Back," *Discover* 20(7, 1999): 76–82.

39: The term "**thermophile**" is used to describe that subgroup of extremophiles that thrive at high temperatures. The first report of a thermophile, discovered in the **River Seine,** was by P. Miquel in *Annals Micrographie* 1(1888): 3–10. The information on the 1926 discovery by **Edson Bastin and Frank Greer** of microbes in oil deposits is from J. Fredrickson and T. C. Onstott, "Microbes Deep Inside the Earth," *Scientific American* 275(1996): 68–73. The subsurface biology research conducted by **Claude Zobell** and Russian scientists during the 1940s and 1950s is described in W. Ghiorse and J. Wilson, "Microbial Ecology of the Terrestrial Subsurface," *Advances in Applied Microbiology* 33(1988): 107–72.

40: The story of the discovery of ***Thermus aquaticus*** is summarized in T. D. Brock, "The Road to Yellowstone—and Beyond," *Annual Review of Microbiology* 49(1995): 1–28.

41: The **EPA subsurface biology program** is described in W. Ghiorse and J. Wilson, "Microbial Ecology of the Terrestrial Subsurface," *Advances in Applied Microbiology* 33(1988): 107–72. The **DOE Subsurface Science Program** is described in J. Fredrickson and T. C. Onstott, "Microbes Deep Inside the Earth," *Scientific American* 275(1996): 68–73. A number of other projects are summarized in R. Kerr, "Life Goes to the Extremes in the Deep Earth—And Elsewhere?" *Science* 276(1997): 703–4.

42: The original paper estimating **the magnitude of the deep, hot biosphere** was Gold's "The Deep, Hot Biosphere," *Proceedings of the National Academy of Sciences USA* 89(1992): 6045–49. Gold has since written a book elaborating on the subject, *The Deep, Hot Biosphere* (New York: Copernicus/Springer-Verlag, 1998). A more detailed quantitative estimate of subsurface biomass is provided in W. Whitman, D. Coleman, and W. Wiebe, "Prokaryotes: The Unseen Majority," *Proceedings of the National Academy of Sciences USA* 95(1998): 6578–83.

43: Information about the **high-temperature record holder** was published in a new journal dedicated to the subject of extremophiles: E. Blöchl et al., "*Pyrolobus fumarii* Represents a Novel Group of Archaea, Extending the Upper Temperature Limit for Life," *Extremophiles* 1(1997): 14–21.

44: The **mechanisms for high-temperature tolerance** remain the subject of much research. Some of what we know today is summarized in R. Atlas and R. Bartha, *Microbial Ecology* (Reading, Mass.: Addison Wesley Longman, 1998), pp. 294–300; and K. O. Stetter, "Hyperthermophiles: Isolation, Classification, and Properties," in *Extremophiles: Microbial Life in Extreme Environments,* edited by K. Horikoshi and W. Grant (New York: Wiley-Liss, 1998), pp. 1–24.

44: **Aerobic and anaerobic respiration** are described in most biology texts, such as in L. Prescott, J. Harley, and D. Klein, *Microbiology* (Boston: William C. Brown, 1996), pp. 165–78.

47: For more on the **Murchison meteorite,** see K. Kvenvolden et al., "Evidence for Extraterrestrial Amino Acids and Hydrocarbons in the Murchison Meteorite," *Nature* 228(1970): 923–26. The fact that this meteorite and others like it contain many of the building blocks of life discussed in chapter 1 suggests yet another hypothesis for the origin-of-life puzzle—that many of the basic ingredients for life did not have to be synthesized on primitive Earth but arrived by meteorite.

49: The original report of the **Columbia River SLiME** is in T. O. Stevens and J. P. McKinley, "Lithoautotrophic Microbial Ecosystems in Deep Basalt Aquifers," *Science* 270(1995): 450–54. Some are skeptical that enough hydrogen is produced in these environments to support microbes, as explained in R. Anderson et al., "Evidence Against Hydrogen-Based Microbial Ecosystems in Basalt Aquifers," *Science* 281(1998): 976–77.

50: The implications of lithotrophic microbes on Earth for the possibility of **subsurface life on other planets** is examined in T. O. Stevens, "Subsurface Microbiology and the Evolution of the Biosphere," in Amy and Haldeman, *The Microbiology of the Terrestrial Deep Subsurface,* pp. 205–23; and P. Boston et al., "On the Possibility of Chemosynthetic Ecosystems in Subsurface Habitats on Mars," *Icarus* 95(1992): 300–308.

50: The original report of the **Martian meteorite** is in D. S. McKay et al., "Search for Past Life on Mars: Possible Relic Biogenic Activity in the Martian Meteorite ALH84001," *Science* 273(1996): 924–30. A popular account is given in D. Goldsmith, *The Hunt for Life on Mars* (New York: Plume Penguin Putnam, 1998). Recent skepticism about signs of life on the Martian rock is summarized in R. Kerr, "Requiem for Life on Mars? Support for Microbe Fades," *Science* 282(1998): 1398–1400.

50: Plans for the **National Astrobiology Institute** are described in J. Wakefield, "The Search for Extreme Life," *Scientific American* 277(7, 2000): 30–31.

51: The quote by **Stephen Jay Gould** about bacteria is from S. J. Gould, "Microcosmos," *Natural History* 105(3): 68.

Chapter 3: Shaking the Tree of Life

53: The complete reference for the **epigraph** is C. Woese, "Default Taxonomy: Ernst Mayr's View of the Microbial World," *Proceedings of the National Academy of Sciences USA* 95(1998): 11044.

55: The **methanogens** use the oxygen in carbon dioxide as their source of oxygen in respiration and release methane (a carbon atom combined with four hydrogen atoms, CH_4) as a by-product. In other words, these anaerobic microbes respire by breathing *in* carbon dioxide and exhaling methane, in contrast to aerobes like ourselves, which breathe in oxygen and *exhale* carbon dioxide. The methanogens are discussed in chapter 9, since the methane they produce is also a so-called greenhouse gas.

56: Although **Woese's office space** was not glamorous, his working laboratory, which we visited later, contained the type of state-of-the-art equipment one would expect to find in a cutting-edge molecular biology research facility.

57: The **"molecular clock"** paper that influenced Woese was E. Zuckerkandl and L. Pauling, "Molecules as Documents of Evolutionary History," *Journal of Theoretical Biology* 8(1965): 357–66.

58: For a relatively easy-to-read explanation of **Carl Woese's methods** (in his own words), see C. Woese, "Archaebacteria," *Scientific American* 244(6, 1981): 98–122. For a more comprehensive and technical explanation, see C. Woese, "Bacterial Evolution," *Microbiological Reviews* 51(2, 1987): 221–71.

60: **Aristotle's "Ladder of Life"** is described in C. Singer, *A Short History of Biology* (Oxford: Clarendon Press, 1931), pp. 39–41.

61: Robert Whittaker proposed his **five-kingdom tree of life** in "New Concepts of Kingdoms of Organisms," *Science* 163(1969): 150–60.

63: **Ralph Wolfe** had earlier co-authored a paper with one of his graduate students describing some general characteristics of this methanogen; see J. Zeikus and R. Wolfe, "*Methanobacterium thermoautotrophicum* sp. n., an Anaerobic, Autotrophic, Extreme Thermophile," *Journal of Bacteriology* 109(1972): 707–13.

64: Woese's publication announcing the **discovery of a new domain** of life was C. R. Woese and G. E. Fox, "Phylogenetic Structure of the Prokaryotic Domain: The Primary Kingdoms," *Proceedings of the National Academy of Sciences USA* 74(1977): 5088–90. Woese's formal proposal of a **three-domain tree of life** was published in C. Woese, O. Kandler, and M. Wheelis, "Towards a Natural System of Organisms: Proposal for the Domains Archaea, Bacteria, and Eucarya," *Proceedings of the National Academy of Sciences USA* 87(1990): 4576–79.

64: For a discussion of the **implications of Woese's universal tree of life,** see N. Pace, "A Molecular View of Microbial Diversity and the Biosphere," *Science* 276(1997): 734–40; and E. Pennisi, "Genome Data Shake Tree of Life," *Science* 280(1998): 672–74.

66: The successful sequencing of the complete genome of *Methanococcus jannaschii* was reported in C. Bult et al., "Complete Genome Sequence of the Methanogenic Archaeon *Methanococcus jannaschii,*" *Science* 273(1996): 1058–73.

66: For a clear, yet thorough, discussion of the problem of **lateral gene transfer** in establishing evolutionary relationships at the base of the universal tree of life, see W. F. Doolittle, "Uprooting the Tree of Life," *Scientific American* 282(2, 2000): 90–95.

67: Discoveries of **archaea in non-extreme habitats** such as common soils are described in K. Jarrell et al., "Recent Excitement About the Archaea," *BioScience* 49(7, 1999): 530–41.

68: The story of **Woese's personal struggle** to gain acceptance of his work was derived from my conversations with him and from the excellent article by Virginia Morell, "Microbiology's Scarred Revolutionary," *Science* 276(1997): 699–702.

69: The story of **Antony van Leeuwenhoek's struggle** for acceptance of his work is described in C. Dobell, *Antony van Leeuwenhoek and His "Little Animals"* (New York: Dover Publications, 1960).

70: For the perspective of **those who are skeptical about Woese's approach** to the study of evolution (and a firsthand look at a scientific debate in progress), I suggest Ernst Mayr's article "Two Empires or Three?" *Proceedings of the National Academy of Sciences USA* 95(1998): 9720–23; followed by Woese's rebuttal, "Default Taxonomy: Ernst Mayr's View of the Microbial World," *Proceedings of the National Academy of Sciences USA* 95(1998): 11043–46. One of the primary arguments of the skeptics is that, at the cellular level, there are basically just two groups of organisms—the prokaryotes (bacteria and archaea), whose cells look very similar to each other, and the eukaryotes (protozoa, fungi, plants, animals), whose cells look similar to each other but very distinct from the prokaryotes. Woese does not quibble with this point but observes that at the *molecular* level there are clearly *three* distinct groups, not just two, and that it is at the molecular level that we can best determine the genealogy of Earth's life forms. Skeptics also argue that a tree of life that lumps together all of the amazing diversity of body shape and behavior we see among the fungi, plants, and animals is simply not very useful. Woese counters that, since the time of Darwin, the goal of classification is not to group organisms in some convenient descriptive fashion but rather to illustrate the true pattern of evolution. For this purpose, Woese and his supporters argue, the rRNA analysis is the best method we have. For now at least.

71: The **microbiology text with Woese's universal tree** inside the front cover is T. D. Brock et al., *Biology of Microorganisms,* 6th ed. (Englewood Cliffs, N.J.: Prentice-Hall, 1991).

Chapter 4: Out of Thin Air

77: Prior to the **evolution of nitrogen-fixing microbes,** some atmospheric N_2 gas was "fixed" by the physical processes of lightning and meteor impacts. The energy released during these events can break apart the nitrogen atoms of N_2 molecules; these nitrogen atoms then combine with oxygen and fall to the Earth in the usable nitrate form. Today, on a global basis, the amount of nitrogen fixed in this way is less than 10 percent of the total, a trickle compared to biological fixation, and not enough to support life on Earth as we know it. For more details on the evolution of the nitrogen cycle, see R. Mancinelli and C. McKay, "The Evolution of Nitrogen Cycling," *Origin of Life and Evolution of the Biosphere* 18(1988): 311–25. See also W. H. Schlesinger, *Biogeochemistry: An Analysis of Global Change* (San Diego: Academic Press, 1997), pp. 32–40; and for a nontechnical version, see T. Volk, *Gaia's Body: Toward a Physiology of Earth* (New York: Copernicus/Springer-Verlag, 1998), pp. 33–44, 221–34.

78: Primary sources of information on **nitrogen fixation and the nitrogenase enzyme** are J. Postgate, *Nitrogen Fixation,* 3d ed. (Cambridge: Cambridge University Press, 1998); Atlas and Bartha, *Microbial Ecology,* pp. 108–12, 418–20; and Prescott et al., *Microbiology,* pp. 199–200.

80: **Virgil** is quoted from *The Georgics,* book I, lines 70–84. This and other material from *The Georgics* was brought to my attention by Logan in *Dirt,* p. 172.

80: The story of **Hellreigel and Wilfarth** and other aspects of the history of the research into symbiotic nitrogen fixation is from P. S. Nutman, "A Century of Nitrogen Fixation Research," *Philosophical Transactions of the Royal Society of London* 317(1987): 69–106.

82: The molecular-level details of the **nodulation process** were derived from Postgate, *Nitrogen Fixation,* pp. 63–95; Atlas and Bartha, *Microbial Ecology,* pp. 112–15; and the recent review by J. Stougaard, "Regulators and Regulation of Legume Root Nodule Development," *Plant Physiology* 124(2000): 531–40.

84: Note that I focus on the **terrestrial nitrogen cycle.** Much is also going on in the oceans, and recent evidence suggests that the amount of nitrogen fixed by free-living cyanobacteria in the oceans may not be that much less than symbiotic fixation on land (Dr. Robert W. Howarth, personal communication, September 2000). For a more comprehensive and quantitative summary of the global nitrogen cycle, see Schlesinger,

Biogeochemistry, pp. 385–95. For an account written for the nonscientist, see Volk, *Gaia's Body.*

86: The concern in the early 1900s that **human demands for fixed nitrogen** were exceeding supply is discussed in C. Delwiche, "The Nitrogen Cycle," *Scientific American* 223(3, 1970): 137–46; and in M. Goran, *The Story of Fritz Haber* (Norman: University of Oklahoma Press, 1967), pp. 42–43.

87: Biographical information on **Fritz Haber** was primarily derived from Goran, *The Story of Fritz Haber,* and from discussions with Dr. Thomas Eisner, whose father was one of Haber's last graduate students.

88: The quote by **Hitler** regarding the dismissal of Jewish scientists is from Goran, *The Story of Fritz Haber,* p. 163.

89: Two excellent papers on the **environmental problems caused by the human fixation of mass quantities of nitrogen** are R. Socolow, "Nitrogen Management and the Future of Food: Lessons from the Management of Energy and Carbon," *Proceedings of the National Academy of Sciences USA* 96(1999): 6001–8; and P. M. Vitousek, "Beyond Global Warming: Ecology and Global Change," *Ecology* 75(7, 1994): 1861–76. A comprehensive assessment of **nitrate pollution of waterways** is included in a recent report by Robert W. Howarth and his colleagues at the National Research Council, *Understanding and Reducing the Effects of Nutrient Pollution* (Washington, D.C.: National Academy Press, 2000).

91: Yet another **leak from applied nitrogen fertilizers** (including manure and other organic fertilizers) is nitrogen that escapes into the atmosphere as ammonia gas rather than being taken up by crops or remaining in the root zone. Ammonia, like nitrous oxide, dissolves into airborne water droplets, acidifying the water, and some of this eventually falls to Earth as acid rain (see also chapter 9).

92: Improving **the efficiency of nitrogen use** in agriculture has been one aspect of my own research and extension program at Cornell since 1984. The information provided here is based primarily on this experience.

Chapter 5: Nexus of the Underground

94: The hypothesis that the **evolution of land plants** was dependent on a mycorrhizal symbiosis with fungi is fully explored in a highly readable popular account by Mark and Dianna McMenamin, *Hypersea: Life on Land* (New York: Columbia University Press, 1994). A summary of recent scientific evidence in support of this hypothesis can be found in M. Blackwell, "Terrestrial Life—Fungal from the Start?" *Science* 289(2000): 1884–85.

94: For a comprehensive summary of plant and fungal species involved in **mycorrhizal associations,** see S. Smith and D. Read, *Mycorrhizal Symbiosis,* 2d ed. (London: Academic Press, 1997). The **ecological**

significance of mycorrhizae is emphasized in M. Allen, *The Ecology of Mycorrhizae* (Cambridge: Cambridge University Press, 1991).

95: The **Rhynie Chert** fossil evidence of ancient mycorrhizal associations is presented in W. Remy et al., "Four Hundred-Million-Year-Old Vesicular-Arbuscular Mycorrhizae," *Proceedings of the National Academy of Sciences USA* 91(1994): 11841–43.

96: Recent **genetic evidence** of the long history of the symbiosis between fungi and land plants is reported in D. R. Redecker et al., "Glomalean Fungi from the Ordovician," *Science* 289(2000): 1920–21; and L. Simon et al., "Origin and Diversification of Endomycorrhizal Fungi and Coincidence with Vascular Land Plants," *Nature* 363(1993): 67–69.

96: We are still learning about the mechanisms by which **plant hosts and mycorrhizal fungi** connect. Recent research suggests that "helper bacteria" in the soil assist in establishing the connection; see J. Garbaye, "Helper Bacteria: A New Dimension to the Mycorrhizal Symbiosis," *New Phytologist* 128(1994): 197–210.

97: For more on the **history of the discovery of mycorrhizae,** see M. C. Rayner, *Mycorrhiza: An Account of Non-Pathogenic Infection by Fungi in Vascular Plants and Bryophytes* (London: Wheldon and Wesley, 1939).

97: For more information on **truffles,** see R. Walsh, "Seeking the Truffle," *Natural History* 105(1, 1996): 20–23; and G. Hudler, *Magical Mushrooms, Mischievous Molds* (Princeton, N.J.: Princeton University Press, 1998), pp. 164–66.

98: **A. B. Frank's** original publication on the mycorrhizae was "Neue Mittheilungen ueber die Mykorrhiza der Baume u. der Monotropa Hypopitys," *Berichte der Deutsche Botanische Gesellschaft* 3(1885): 27–40.

99: **Gallaud's drawings** of the microscopic arbuscular mycorrhizae appeared in "Etudes sur les mycorrhizes endotrophs," *Revue Générale de Botanique* 17(1905): 1-48, Plates 1-4.

99: The theory that **roots evolved from the ancient algal-fungus symbiosis** was proposed by K. Pirozynske and D. Malloch, "The Origin of Land Plants: A Matter of Mycotropism," *BioSystems* 6(1975): 153–64.

99: For a clear explanation of the **endosymbiosis theory** of evolution, in her own words, see Lynn Margulis, "Symbiosis and Evolution," *Scientific American* 225(2, 1971): 49–57.

102: The data on the **length of mycorrhizal hyphae** in soil are from Allen, *The Ecology of Mycorrhizae*, p. 25.

102: The **short-circuiting of the nutrient cycle** by mycorrhizae is discussed in Atlas and Bartha, *Microbial Ecology*, p. 107.

102: The first experiment to show clearly **plant-to-plant nutrient transfer through mycorrhizal connections** was F. Woods and K. Brock, "Interspecific Transfer of Ca-45 and P-32 by Root Systems," *Ecology* 45(4, 1964): 886–89. One of many observations of nitrogen transfer from a legume to a nonlegume plant was made by G. Bethlenfalvay et al., "Nutrient Transfer Between the Root Zones of Soybean and Maize Plants Connected by a Common Mycorrhizal Mycelium," *Physiologia Plantarum* 82(1991): 423–32. E. I. Newman wrote a useful review, "Mycorrhizal Links Between Plants: Their Functioning and Ecological Significance," *Advances in Ecological Research* 18(1988): 243–70. An update of recent research is found in S. Simard et al., "Net Transfer of Carbon Between Mycorrhizal Tree Species in the Field," *Nature* 388(1997): 579–82.

104: The **220,000-pound fungus** is described in M. Smith et al., "The Fungus *Armillaria bulbosa* Is Among the Largest and Oldest Living Organisms," *Nature* 356(1992): 428–31.

104: The **skepticism about the importance of mycorrhizae** shared by many scientists until relatively recently is reviewed by Michael Allen in his introduction to *The Ecology of Mycorrhizae,* pp. 1–3.

105: The quote by **Richard Dawkins** is from R. Dawkins, *The Selfish Gene,* 2d ed. (Oxford: Oxford University Press, 1989), p. 233.

Chapter 6: When the Humble Explain the Great

107: Adrian Desmond and James Moore, in *Darwin,* their biography of the famous scientist (New York: W. W. Norton, 1991), comment: "For him [Darwin] **the humble explained the great**" (p. 657), referring to his lifelong interest in the "humble" earthworm. Thus, the title of this chapter.

107: The depiction of **Darwin's study-laboratory,** the experiments in progress, and his poor state of health are derived from Desmond and Moore, *Darwin,* and from C. Darwin, *The Formation of Vegetable Mould Through the Action of Worms, with Observations of Their Habits* (London: Murray, 1881).

110: Darwin's 1837 visit with his uncle is described as **the beginning of his interest in earthworms** by O. Graff, "Darwin on Earthworms—The Contemporary Background," in *Earthworm Ecology: From Darwin to Vermiculture,* edited by J. E. Satchell (New York: Chapman and Hall, 1983), pp. 5–18.

110: Originally published in 1881, *The Formation of Vegetable Mould Through the Action of Worms* is available in a more recent edition: *Darwin on Earthworms* (London: Bookworm Publishing, 1976), and it also appears in some compilations of Darwin's writings.

112: The proceedings of the centenary **international symposium on earthworms** were published as Satchell, *Earthworm Ecology.* There have been numerous international earthworm conferences since then—for example, the proceedings of the Fifth International Symposium on Earthworm Ecology, edited by C. A. Edwards, were published in *Soil Biology and Biochemistry* (special issue) 29(3, 1997): 215–750.

113: The **reproductive biology and respiration** of earthworms are described in C. A. Edwards and P. J. Bohlen, *Biology and Ecology of Earthworms,* 3d ed. (London: Chapman and Hall, 1996), pp. 55–60, 71–72.

116: Darwin describes his research on **earthworm digestion and sensory capabilities** in *The Formation of Vegetable Mould Through the Action of Worms,* pp. 19–35. The anecdotes about how **Darwin got the family involved** in his worm research are from Desmond and Moore, *Darwin,* p. 649.

116: For more information on **earthworm species distribution** worldwide, earthworm behavior, and longevity, see Edwards and Bohlen, *Biology and Ecology of Earthworms.*

117: **Darwin's long-term field experiment** to document the earth-moving capabilities of worms, and his work at Stonehenge, are explained in detail in *The Formation of Vegetable Mould Through the Action of Worms,* pp. 139–42, 154–56. **Darwin on the importance of worms** is quoted from p. 313.

118: For more information about **termites,** see M. Wood, *Environmental Soil Biology* (London: Chapman and Hall, 1995), pp. 80–82. For information about **ants,** I highly recommend the Pulitzer Prize–winning *The Ants* by Bert Hölldobler and Edward O. Wilson (Cambridge, Mass.: Belknap Press of Harvard University Press, 1990).

118: Regarding the interactions between earthworms and microbes, the importance of **protozoans in the diet** of earthworms was reported in H. Miles, "Soil Protozoa and Earthworm Nutrition," *Soil Science* 95(1963): 407–9. The beneficial effect of earthworms on the **spread of apple scab** was noted in J. Hirst et al., "The Origin of Apple Scab in the Wisbech Area in 1953 and 1954," *Plant Pathology* 4(1955): 91.

119: The importance of earthworms in nutrient cycling, **soil formation,** and improving soil quality is discussed in detail in Edwards and Bohlen, *Biology and Ecology of Earthworms;* see also Coleman and Crossley, *Fundamentals of Soil Ecology,* pp. 98–105; and Wood, *Environmental Soil Biology,* pp. 26–28.

120: The **Canadian fish bait industry** is described in A. D. Tomlin, "The Earthworm Bait Market in North America," in Satchell, *Earthworm Ecology,* pp. 331–38.

121: The use of earthworms in agriculture and **land reclamation** is described in Edwards and Bohlen, *Biology and Ecology of Earthworms;* and in K. E. Lee, *Earthworms: Their Ecology and Relationships with Soils and Land Use* (Sydney: Academic Press, 1985).

122: The **dinner scene at the Darwin home** occurred on September 28, 1881, as described in Desmond and Moore, *Darwin,* pp. 656–58. The visitors were the political reformers Edward Aveling and Ludwig Büchner. Darwin's lifelong project—his book on worms—appeared and became a bit of a sensation just a couple of weeks after this visit. Darwin died the following year, on April 19, 1882.

Chapter 7: Germ Warfare

123: The **epigraph** is from a speech to the National Academy of Sciences, as quoted in S. Waksman, *My Life with the Microbes* (New York: Simon and Schuster, 1954), p. 11.

123: *Streptococcus pyogenes* and other *Streptococcus* **species that live on the skin** can cause the so-called flesh-eating disease. Some of these appear to be increasing their resistance to antibiotics. *Staphylococcus* **species** can cause pimples, boils, and other pus-forming infections, even without a wound site for entry. For more information, see Prescott et al., *Microbiology,* pp. 746–48, 759–62.

124: The **Edwin Smith Papyrus, Hippocrates' observations,** and other historical information on tetanus can be found in D. Guthrie, *A History of Medicine,* 2d ed. (London: Thomas Nelson and Sons, 1958), pp. 20, 59.

124:/ For more information on the **symptoms, epidemiology, and treatment of tetanus,** see F. E. Udwadia, *Tetanus* (Bombay: Oxford University Press, 1994). I also used as references Prescott et al., *Microbiology,* pp. 764–66; and the website www.WebMD.com.

125: The development of the vaccine for tetanus by **von Behring and Kitasato** is described in P. Baldry, *The Battle Against Bacteria* (Cambridge: Cambridge University Press, 1976), pp. 72–73.

126: **Infant mortality from tetanus** is discussed in Udwadia, *Tetanus,* pp. 9–17. For current statistics, see the website www.WHO.INT/vaccines-diseases/diseases/neonataltetanus.

126: For more information on **gas gangrene** and **soilborne fungal diseases of humans,** see Prescott et al., *Microbiology,* pp. 756, 791–97; and Hudler, *Magical Mushrooms, Mischievous Molds,* pp. 99–112.

128: For a comprehensive treatment of **soilborne plant diseases,** see C. H. Dickinson and J. A. Lucas, *Plant Pathology and Plant Pathogens* (Oxford: Blackwell, 1982).

129: A. Bourke provides an excellent reexamination of the **Irish potato famine** story in *The Visitation of God? The Potato and the Great Irish Famine* (Dublin: Lilliput Press, 1993). The **reemergence of potato late blight** as a major threat to farmers and national economies is explained in W. E. Fry and S. B. Godwin, "Resurgence of the Irish Potato Famine Fungus," *BioScience* 47(6, 1997): 363–71; and D. Douglas, "The Leaf That Launched a Thousand Ships," *Natural History* 105(1, 1996): 24–32. Some of my information came from personal communications with Dr. William Fry, a leading authority on *Phytopthora infestans.*

132: The suppression of **papaya root rot** by introducing topsoil from another site was reported in M. Schroth and J. Hancock, "Selected Topics in Biological Control," *Annual Reviews of Microbiology* 35(1981): 453–76. The same authors published a paper that stimulated much interest a year later: "**Disease-Suppressive Soil** and Root-Colonizing Bacteria," *Science* 216(1982): 1376–81. Also see H. Hoitink and M. Boehm, "Biocontrol within the Context of Soil Microbial Communities," **Annual Review of Phytopathology** 37 (1999): 427-46

133. An excellent recent review of *Trichoderma* research and a general discussion of the prospects for biocontrol of plant diseases are provided in G. Harman, "Myths and Dogmas of Biocontrol: Changes in Perception Derived from Research on *Trichoderma harzianum* T-22," *Plant Disease* 84(4, 2000): 377–92.

136: The chronology of **Selman Waksman's research** that led to the discovery of streptomycin is provided in his autobiography: Waksman, *My Life with the Microbes.*

Chapter 8: Endangered Diggers of the Deep

143: The epigraph quote by **David Wilcove** of Environmental Defense is from A. Dobson and A. Lyles, "Black-Footed Ferret Recovery," *Science* 288(2000): 988.

143: My primary source for the account of **Lewis and Clark's encounters with prairie dogs** is the Thwaites edition of the journals: *Original Journals of the Lewis and Clark Expedition,* vol. 1, edited by R. G. Thwaites (New York: Antiquarian Press, 1959). Secondary sources and background information come from S. Ambrose, *Undaunted Courage* (New York: Simon and Schuster, 1996); and R. D. Burroughs, *The Natural History of the Lewis and Clark Expedition* (Lansing: Michigan State University Press, 1961).

146: The **research on prairie dog communication** is reported in C. N. Slobodchikoff et al., "Semantic Information Distinguishing Individual Predators in the Alarm Calls of Gunnison's Prairie Dogs," *Animal Behavior* 42(1991): 713–19.

146: **Zebulon Montgomery Pike's** encounters with prairie dogs are described in E. Coues and J. A. Allen, *Monographs of North American Rodentia: U.S. Survey of the Territories,* vol. 11 (Washington, D.C.: U.S. Government Printing Office, 1877), pp. 889–909. **General Custer's** journal entries regarding prairie dogs are described in E. Andersen et al., "Paleobiology, Biogeography, and Systematics of the Black-footed Ferret," *Great Basin Naturalist Memoirs* 8(1986): 11–62.

147: The **25,000-square-mile prairie dog colony** is described in R. S. Hoffmann, "Black-tailed Prairie Dog (*Cynomys ludovicianus*)," in *The Smithsonian Book of North American Mammals,* edited by D. Wilson and S. Ruff (Washington, D.C.: Smithsonian Institution Press, 1999), pp. 445–47.

147: The ecological role of **prairie dogs as landscape managers** is described in J. L. Hoogland, *The Black-Tailed Prairie Dog* (Chicago: University of Chicago Press, 1995); and in B. Miller, R. Reading, and S. Forrest, *Prairie Night: Black-footed Ferrets and the Recovery of Endangered Species* (Washington, D.C.: Smithsonian Institution Press, 1996).

148: For many more details about the social life and behavior of prairie dogs, see Hoogland, *The Black-tailed Prairie Dog*. A brief popular account is provided in J. Ferrara, "Prairie Home Companion," *National Wildlife* 23(3, 1985): 49–53.

150: For more information on the bacterial plague (also called sylvatic or bubonic plague) and prairie dogs, see Hoogland, *The Black-tailed Prairie Dog*, p. 80.151.

151: A primary source of information on the prairie dog poisoning campaigns of the early 1900s is W. R. Bell, "Death to the Rodents," in U.S. Department of Agriculture, *USDA Yearbook 1920* (Washington, D.C.: U.S. Government Printing Office, 1920), pp. 421–38. The role played by C. H. Merriam and other government officials is described in Miller et al., *Prairie Night*, pp. 22–25.

152: The cost-benefit analysis revealing that poisoning programs operate at a net loss is in A. R. Collins et al., "An Economic Analysis of Black-tailed Prairie Dog Control," *Journal of Range Management* 37(1984): 358–61.

152: Another rationale for the poisoning of prairie dogs has been injury to livestock caused by stepping into burrow entrances. While this can happen, it is probably rare. A recent informal survey was unable to come up with a single verifiable incident of such injury (Hoogland, *The Black-Tailed Prairie Dog*, pp. 20, 21).

153: For the latest information on efforts to protect prairie dog habitat, see the websites for the National Wildlife Federation (www.nwf.org) and the Predator Conservation Alliance (www.predatorconservation.org).

154: For more information on the behavior and ecology of the black-footed ferret, and the story of the battle to save them from extinction, see Miller et al., *Prairie Night*. Some aspects of the story are also told in R. M. May, "The Cautionary Tale of the Black-footed Ferret," *Nature* 320(1986): 13–14; and D. Weinberg, "Decline and Fall of the Black-footed Ferret," *Natural History* 95(2, 1986): 63–69. A recent update of the ferret's status is provided in A. Dobson and A. Lyles, "Black-footed Ferret Recovery," *Science* 288(2000): 985–88. I also relied on personal communications with Dr. Dean Biggins, a U.S. Geological Survey scientist involved in ferret conservation. For the latest information on the status of this rare mammal, see the Predator Conservation Alliance website: www.predatorconservation.org.

157: One of the Zuni Indian stories about the burrowing owl is told in V. C. Holmgren, *Owls in Folklore and Natural History* (Santa Barbara, Calif.: Capra Press, 1988), pp. 52–54.

158: A classic study on the behavior and ecology of the burrowing owl is L. Thomsen, "Behavior and Ecology of Burrowing Owls on the Oakland Municipal Airport," *The Condor* 73(1971): 177–92. A more general reference, including geographical distribution, is E. A. Haug et al., "Burrowing Owl," in American Ornithologists Union, *Birds of North*

America, vol. 61 (Philadelphia: Academy of Natural Sciences of Philadelphia, 1993). A less technical discussion is provided in G. Green, "Living on Borrowed Turf," *Natural History* 97(9, 1988): 58–64.

158: The first description of burrowing owls by Father Molina in 1782 was mentioned in Holmgren, *Owls in Folklore and Natural History,* pp. 154, 155.

161: A recent study documenting the effects of agricultural pesticides on burrowing owls is by J. A. Gervais et al., "Burrowing Owls and Agricultural Pesticides: Evaluation of Residues and Risks for Three Populations in California," *Environmental Toxicology and Chemistry* 19(2, 2000): 337–43.

162: Various aspects of the struggle to protect burrowing owl habitat in the Silicon Valley are described in R. Holmes, "City Planning for Owls," *National Wildlife* (October-November 1998): 46–53; and L. Trulio, "Native Revival: Efforts to Protect and Restore the Burrowing Owl in the South Bay," *Tideline* [a publication of the U.S. Department of the Interior, Don Edwards San Francisco Bay National Wildlife Refuge] 18(1, 1998): 1–3. Additional information was obtained in personal communications with Dr. Lynne Trulio.

163: Mitigation efforts to help owls survive on very urbanized sites are described in L. Trulio, "Passive Relocation: A Method to Preserve Burrowing Owls on Disturbed Sites," *Journal of Field Ornithology* 66(1, 1994): 99–106; and L. Trulio, "Burrowing Owl Demography and Habitat Use at Two Urban Sites in Santa Clara County, California," *Journal of Raptor Research Reports* 9(1997): 84–89.

Chapter 9: The Good Earth

165: The estimate that 97 percent of our food is obtained from the soil (and 3 percent from the oceans and other aquatic sources) is from the Council on Environmental Quality (CEQ), *The Global 2000 Report to the President,* vol. 2 (Washington, D.C.: U.S. Government Printing Office, 1980).

166: The "worried ecologist" quote by Edward O. Wilson is from E. O. Wilson, *Consilience: The Unity of Knowledge* (New York: Alfred A. Knopf, 1998), p. 287.

166: The tragic events of the Dust Bowl, as well as its causes and long-term consequences, are described vividly in the excellent book by D. R. Hurt, *The Dust Bowl: An Agricultural and Social History* (Chicago: Nelson-Hall, 1981). See pages 51–52 for a description of the black blizzard of March 15, 1935, with original sources cited.

168: The farmer who claimed that "you just can't seriously hurt this land" is quoted in Hurt, *The Dust Bowl,* p. 68.

168: The rainfall data for Cimarron County, Oklahoma, are from Hillel, *Out of the Earth,* p. 162.

171: The range of **current soil erosion rates for the United States and Europe** are based primarily on data from L. K. Lee, "The Dynamics of Declining Soil Erosion Rates," *Journal of Soil and Water Conservation* 45(1990): 622–44; and D. Pimentel and N. Kounang, "Ecology of Soil Erosion in Ecosystems," *Ecosystems* 1(1998): 416–26. A much lower estimate of soil erosion losses, based on rates of sediment deposition into waterways, is suggested by S. Trimble and P. Crosson, "U.S. Soil Erosion Rates—Myth and Reality," *Science* 289(2000): 248–50.

171: **Rates of soil formation** (usually hundreds of years per inch of new topsoil) are discussed in many soil science texts, such as Jenny, *The Soil Resource.*

171: Rates of **soil erosion in the tropics,** and many relevant references, are provided in *World Soil Erosion and Conservation,* edited by D. Pimentel et al. (Cambridge: Cambridge University Press, 1993). This topic is also discussed in Hillel, *Out of the Earth,* pp. 163–65, 205–10.

172: For off-site **costs of erosion** and the debate about economic impacts, see D. Pimentel et al., "Environmental and Economic Costs of Soil Erosion Rates and Conservation Benefits," *Science* 267(1995): 1117–23; and the opposing, much more optimistic, view by P. Crosson, "Soil Erosion Costs and Estimates," *Science* 269(1995): 461–63.

173: Modern **recommendations for soil conservation,** written for farmers and home gardeners, are provided in F. Magdoff and H. van Es, *Building Soils for Better Crops* (Burlington, Vt.: Sustainable Agriculture Publications, 2000).

174: The most comprehensive text on the use of **soil microbes for bioremediation** of toxic sites is Martin Alexander, *Biodegradation and Bioremediation,* 2d ed. (San Diego: Academic Press, 1999). The **white-rot fungi** are described on p. 349, and genetic engineering of *Pseudomonas putida* is discussed on p. 318. Another useful reference is Atlas and Bartha, *Microbial Ecology,* pp. 512–85.

178: For the connection between **acid rain and forest decline** in Europe and the history of research on this topic, see E.-D. Schulze, "Air Pollution and Forest Decline in a Spruce (*Picea abies*) Forest," *Science* 244(1989): 776–83.

178: The effect of **nitrogen deposition** (associated with acid rain) and forest decline (in the northeastern United States) is discussed in J. Aber et al., "Nitrogen Saturation in Temperate Forest Ecosystems," *BioScience* 48(11, 1998): 921–34. Other aspects of the effects of acid rain on plant nutrient availability are explored in W. Shortle and K. Smith, "Aluminum-Induced Calcium Deficiency Syndrome in Declining Red Spruce," *Science* 240(1988): 1017–18.

179: Eef Arnolds's comprehensive study of the **decline of soil mycorrhizal fungi in central Europe** and the connection with the decline in the health of evergreen tree species is summarized in his paper "Decline of

Ectomycorrhizal Fungi in Europe," *Agriculture, Ecosystems, and Environment* 35(1991): 209–44.

179: The success of the 1990 Clean Air Act at **reducing sulfate emissions** in the United States is described in R. Kerr, "Acid Rain Control: Success on the Cheap," *Science* 282(1998): 1024–27.

180: The **effect of soil biological processes on greenhouse gases** has been the focus of much recent research and a number of international scientific conferences. For a sampling, see *Soils and the Greenhouse Effect*, edited by A. Bouwman (Chichester, Eng.: John Wiley and Sons, 1990); and *Soils and Global Change*, edited by R. Lal et al. (Boca Raton, Fla.: Lewis Publishers, 1995).

182: The **effects of carbon dioxide on soil organisms** and nutrient cycling are reviewed in E. Patterson et al., "Effect of Elevated CO_2 on Rhizosphere Carbon Flow and Soil Microbial Processes," *Global Change Biology* 3(1997): 363–77; and M. Sadowsky and M. Schortemeyer, "Soil Microbial Responses to Increased Concentrations of Atmospheric CO_2," *Global Change Biology* 3(1997): 217–24. At Cornell, I am collaborating with the soil microbiologist Janice Thies to evaluate the effects of CO_2 on soil biodiversity, using molecular genetic techniques, and the effects on nutrient availability to plants. Thus far, we have shown (as have other groups) that symbiotic nitrogen fixation is stimulated in legume plants grown at higher carbon dioxide levels (because photosynthesis is stimulated and the plants can support a higher population of bacterial symbionts on their roots). Higher carbon dioxide levels will thus benefit nitrogen-fixing weed species as well as crop species and undoubtedly lead to changes in the mix of plant species and nutrient cycling in natural plant communities. This demonstration is just one example of the complex interaction between above- and below-ground responses to carbon dioxide.

183: The potential for **increasing carbon storage of soils** is explored in R. Lal et al., *The Potential of U.S. Cropland to Sequester Carbon and Mitigate the Greenhouse Effect* (Chelsea, Mich.: Ann Arbor Press, 1998).

185: The leading authority on **bioprospecting**—the chemical ecologist Tom Eisner—provides a particularly lucid explanation of the value of species diversity in "The Hidden Value of Species Diversity," *BioScience* 42(8, 1992): 578.

185: Daniel Hillel discusses **conditional optimism** in *Out of the Earth*, pp. 276–83.

Epilogue

187: The first edition of Walt Whitman's *Leaves of Grass* was published in 1855. The epigraph and other Whitman quotes are from the 1993 Random House edition.

Index

Abawi, George: on sudangrass, 132
Acid rain, 91, 197
  impact of, 176, 178–79
  nitrogen and, 178, 206
  nitrous oxide and, 179
  plant nutrients and, 206
  pollution and, 165–66
Actinomycetes, 138
Adama, defined, 25
Adenosine triphosphate (ATP), 23,
  29, 45, 46
Aerobic organisms, oxygen and, 45
Agaricus brunnescens, 97
Agriculture
  expansion in, 167, 179
  nitrogen fertilizers and, 178
  research and development in,
  174
A horizon, 10, 11
Alfisol, 8
Algae, fungi and, 94, 199
Allen, Michael
  on mycorrhizae, 101
  truffles and, 98
Amaranthaceae, 99
Amino acids, 22, 23, 26, 28, 31, 49,
  57, 75
  changes in, 58
  nitrogen and, 77
  nucleotide synthesis with, 24
  production of, 25
Ammonia, 21, 23, 24, 178, 197

Ammonium, 78, 83, 84, 85
  commercial production of, 88
  nitrate and, 91
  nitrogen gas and, 86, 87–88
Anaerobes, 45–46
  metabolic rate of, 47
  sugar and, 45
And Thus Spake Zarathustra
  (Nietzsche), quote from, 1
Animalcules, 69, 71
Antibiotics, 13, 135, 136, 185
  development of, 137–39
  in soil, 137, 138, 188
  tetanus, 124
Ants, 7 (fig.), 200
Apple scab, earthworms and, 119,
  201
Arbuscular mycorrhizae, 95, 96, 99,
  199 (fig.)
  microscopic images of, 95 (fig.)
Archaea, 64, 75, 77, 175, 180, 196
  domain of, 65–66
  ecological role of, 67
  evolutionary tree and, 70
  in non-extreme habitats, 195
  temperatures for, 44, 67
Archaebacteria, 55
Aristotle, Ladder of Life by, 60, 61
  (fig.), 194
Armillaria bulbosa, 103
Arnolds, Eef: on mycorrhizal fungi,
  179, 206

Forests, decline of, 176–78, 206
Formation of Vegetable Mould Through the Action of Worms, The
(Darwin), 107, 110, 200
Fossil fuels, 179
Fowler, William: on origin of elements, 21
Frank, A. B.
ectomycorrhizae and, 99
mycorrhizae and, 98, 198
root fungi and, 98, 107
truffles and, 97
Fry, William: *Phytopthora infestans* and, 202
Fumigacin, 138
Fungal symbionts, 99
Fungi, 7 (fig.), 10–13, 20, 64, 128, 188
algae and, 94
harmful, 131
infections from, 126, 127
kingdom for, 61
large, 199
mycorrhizal associations with, 105
nonmycorrhizal types of, 103–4
plants and, 93, 98
root, 94, 96, 101, 107, 118, 133
subsurface, 42, 101
symbiosis with, 104
white-rot, 175, 206
Fungicides, 130
Fusarium, 128

Gaia's Body: Toward a Physiology of Earth (Volk), quote from, 75
Galileo, 69
Gallaud, I.: drawings of, 99, 199
Gas gangrene, 126, 202
Genes
clay and, 30, 32
lateral transfer of, 195
low-tech, 26
molecular nature of, 29
RNA and, 57

self-replication of, 30
synthesis of, 25
Genetic codes, tracing, 2, 24, 33, 57, 71
Genetic material, 76, 198
nucleotide sequence of, 53, 58
transferring, 66
"Genetic takeover" hypothesis, 31
Genus-species classification system, 2, 6, 60, 61
Georgics, The (Virgil), 79, 196
Geosmin, actinomycetes and, 138
Global warming, 180
carbon storage and, 183–84
organic matter and, 183
Gold, Thomas
on subsurface biomass, 42–43
deep, hot biosphere and, 43
microbial activity and, 39
Good Earth, The (Buck), quote from, 165
Gould, Stephen Jay, 51, 194
Grapes of Wrath, The (Steinbeck), 168
Grasslands
out of balance, 152
overgrazing of, 167
prairie dogs and, 143, 151
preservation of, 164, 170
Gravitation, increase in, 18
Great Plains
black blizzards on, 169
drought cycles in, 167
prairie dogs and, 151, 153–54
rainfall in, 168
settlers of, 165, 166, 168
Greenhouse gases, 91, 179, 180, 182, 194
formation of, 183
impact of, 176
soil biological processes and, 206–7
soil organisms and, 181 (fig.)
Green manures, 82, 92
Greer, Edson, 192